THE SCHWARZ FUNCTION
AND ITS APPLICATIONS

By

PHILIP J. DAVIS

THE

CARUS MATHEMATICAL MONOGRAPHS

Published by
THE MATHEMATICAL ASSOCIATION OF AMERICA

———

THE CARUS MATHEMATICAL MONOGRAPHS are an expression of the desire of Mrs. Mary Hegeler Carus, and of her son, Dr. Edward H. Carus, to contribute to the dissemination of mathematical knowledge by making accessible at nominal cost a series of expository presentations of the best thoughts and keenest researches in pure and applied mathematics. The publication of the first four of these monographs was made possible by a notable gift to the Mathematical Association of America by Mrs. Carus as sole trustee of the Edward C. Hegeler Trust Fund. The sales from these have resulted in the Carus Monograph Fund, and the Mathematical Association has used this as a revolving book fund to publish the succeeding monographs.

The expositions of mathematical subjects which the monographs contain are set forth in a manner comprehensible not only to teachers and students specializing in mathematics, but also to scientific workers in other fields, and especially to the wide circle of thoughtful people who, having a moderate acquaintance with elementary mathematics, wish to extend their knowledge without prolonged and critical study of the mathematical journals and treatises. The scope of this series includes also historical and biographical monographs.

The following monographs have been published:

QA 331
.D355

THE SCHWARZ FUNCTION
AND ITS APPLICATIONS

By
PHILIP J. DAVIS
Professor of Applied Mathematics
Brown University

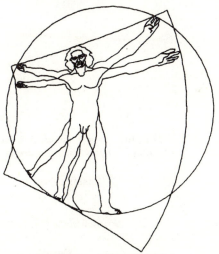

Published and Distributed by
THE MATHEMATICAL ASSOCIATION OF AMERICA

© *1974 by*

The Mathematical Association of America (Incorporated)
Library of Congress Catalog Card Number 74-77258

Complete Set ISBN 0-88385-000-1
Vol. 17 ISBN 0-88385-017-6

Printed in the United States of America

Current printing (last digit):

10 9 8 7 6 5 4 3 2 1

To

Stefan and Edy Bergman

In Friendship

CONTENTS

ACKNOWLEDGEMENTS

I should like to acknowledge my pleasure at the encouragement provided me by the Committee on Carus Monographs.

I am indebted to the following individuals: Eleanor M. Addison, Katrina Avery, Maylun Buck-Lew, Ezoura Fonseca, Frances Gajdowski, Carol Kemmler, Carol Salvatore, Dr. Henry O. Pollak, Professor J. H. Ahlberg, Professor Martin Braun, Professor Daniel Finkbeiner, Professor Peter Henrici, Professor Herbert Kolsky, Professor Philip Rabinowitz and Professor Frank Stenger.

The motif on the title page is the da Vinci rendition of the human figure according to the proportions of Vitruvius. It has been subjected to a Möbius transformation leaving the bounding circle invariant. This was kindly provided to me by Professors R. Vitale of Brown and K. Long of the Rhode Island School of Design. The end piece was created by Jonathan Sachs.

The manuscript was prepared partially under the support of the National Science Foundation under Grant GP-35398 with Brown University.

PHILIP J. DAVIS

Brown University

CHAPTER 1

PROLOGUE

In the year 1968–1969, Professor Mary Cartwright was a visiting member of the Division of Applied Mathematics at Brown University. This was a year of turmoil at Brown—particularly curricular turmoil—and in the course of one of our division meetings Miss Cartwright remarked that when she was a student all mathematics majors were required to know a proof of the nine-point circle theorem. Since the nine-point circle now has a distinct flavor of beautiful irrelevance, Miss Cartwright seemed to be telling us that we should not be too dogmatic as to what constitutes a proper mathematics curriculum. Fashion is spinach even in mathematics, and time often works to "nine-point circle-ize" many of our most relevant and sophisticated topics that are now insisted upon.

At the time, I was giving a course in Numerical Analysis entitled "Iteration Theory in Banach Spaces", using notes of L. B. Rall, and I saw that it would be possible—and not too far-fetched—to present several lectures which would trace an unlikely path from the nine-point circle to iteration.

This essay presents such a path. The connecting link is the use of conjugate coordinates and the Schwarz reflection function. The path has been faired, as draftsmen say, to pass in a wide arc near a number of allied topics in complex variable theory that have interested me.

In Chapter Two conjugate coordinates are introduced. In Chapter Three elementary notions of plane analytic geometry are expressed in terms of conjugate coordinates. In Chapter Four this mechanism is used to establish the nine-point circle theorem and Feuerbach's theorem.

In Chapter Five, the Schwarz Function of an analytic arc is defined and numerous examples are given. The remainder of the book is devoted to expounding applications to and connections with a variety of topics, principally in the theory of analytic functions of a complex variable.

Chapter Six develops the relationship of the Schwarz Function to Schwarzian reflection and an important functional identity is adduced. Chapter Seven turns to elementary plane differential geometry and finds relationships between the differential invariants (curvature, etc.), the Schwarz Function and the Schwarzian derivative. Chapter Eight talks about conformal mapping, reflections and symmetry in analytic arcs and invariant curves. A convenient algebraic formalism involving the Schwarz Function is introduced to handle some of the problems and this leads to a number of interesting functional equations.

Chapter Nine relates the Schwarz Function to 2×2 systems of autonomous differential equations. In Chapter Seven, the Schwarz Function is discussed in the small; Chapter Ten takes up the properties in the large, specifically the circumstances when the Schwarz Function is rational.

Chapter Eleven begins by expressing the operators of partial differential equations in terms of conjugate coordinates. Applications are then made to problems of analytic continuation of harmonic functions satisfying nonlinear boundary data as well as to the Cauchy problem for elliptic equations. In the second part of the chapter, passing from derivatives to integrals, Green's theorem, in its complex analytic form, is used to derive conveniently a number of integral identities, some of which are of interest in the theory of approximate quadratures.

Chapter Twelve shows how a number of ideas and principles of two-dimensional flow theory are easily expressed in terms of Schwarz Functions, while Chapter Thirteen relates the solution of the Dirichlet Problem to the Schwarz Function of the boundary of the region in question.

Chapter Fourteen takes up once again properties in the large, dealing specifically with closed curves whose Schwarz Function is meromorphic inside the curve. Further interesting identities and functional equations are derived, some of which connect up with the Bergman Kernel Function.

In the final chapter, analytic functions of one and two real variables are expressed in conjugate coordinates and certain functional iterations suggested by the Schwarz Function are taken up. The Schroeder function is found to fit into this theory nicely and a number of classic problems, including the "function-theoretic center problem" and the "bisection problem for curvilinear angles", are discussed from this point of view.

I have resisted the temptation to expound such things as the integral operator method for the solution of partial differential equations. Conjugate coordinates are crucial in the operator method, but the Schwarz Function plays a minor role. I have also avoided an extensive reworking of inversive geometry and the Poincaré model of non-

Euclidean geometry in the present terms, feeling that these topics are adequately covered from numerous points of view in the text-book literature.

H. A. Schwarz showed us how to extend the notion of reflection in straight lines and circles to reflection in an arbitrary analytic arc. Notable applications were made to the symmetry principle and to problems of analytic continuation. Reflection, in the hands of Schwarz, is an anti-analytic mapping. By taking its complex conjugate, we arrive at an analytic function that we have called here the Schwarz Function of the analytic arc. This function is worthy of study in its own right and this essay presents such a study. In dealing with certain familiar topics, the use of the Schwarz Function lends a point of view, a clarity and elegance, and a degree of generality which might otherwise be missing. I have also found that it opens up a line of inquiry which has yielded numerous interesting things in complex variables; it illuminates some functional equations and a variety of iterations which interest the numerical analyst. The perceptive reader will certainly find here some old wine in relabelled bottles. But one of the principles of mathematical growth is that the relabelling process often suggests a new generation of problems. Means become ends; the medium rapidly becomes the message.

This book is not wholly self-contained. The reader will find that he should be familiar with the elementary portions of linear algebra and of the theory of functions of a complex variable.

CHAPTER 2

CONJUGATE COORDINATES
IN THE PLANE

Plane Euclidean geometry and, in particular, many of the topics which normally appear in advanced synthetic or inversive geometry may be expeditiously carried out by working with complex coordinates. The real plane is converted into the complex plane by assigning the complex number $z = x + iy$, $i = \sqrt{-1}$, to the real point (x, y). If we wish to recover x and y from z, it is convenient to introduce the conjugate quantity $\bar{z}$ by means of the equations

$$(2.1) \qquad z = x + iy, \qquad \bar{z} = x - iy,$$

and hence the inverse transformation is given by

$$(2.2) \qquad x = \tfrac{1}{2}(z + \bar{z}), \qquad y = \frac{1}{2i}\,(z - \bar{z}).$$

The non-independent quantities z, $\bar{z}$ are called the *conjugate coordinates* of the point (x, y). (If x and y are themselves complex, then z and $\bar{z}$ are independent but are not necessarily conjugate.) In somewhat different contexts involving the geometry of the "complex domain" in which x and y may take complex values, the terms *isotropic* or *minimal* coordinates are used for conjugate coordinates.

5

In this book, *we restrict ourself almost entirely to real values of x and y*, i.e., to the usual Gauss plane.

The matrix of the transformation (2.1) is

$$(2.3) \qquad M = \begin{pmatrix} 1 & i \\ 1 & -i \end{pmatrix}.$$

We have

$$(2.4) \qquad MM^* = 2I,$$

where * designates the conjugate transposed matrix, so that $1/\sqrt{2}M$ is unitary. For some purposes, the corresponding transformation may be somewhat more convenient; indeed, this transformation is widely used in the theory of complex manifolds and of integral operators for the solution of partial differential equations.

CHAPTER 3

ELEMENTARY GEOMETRIC FACTS

Given two distinct points z_1 and z_2 in the complex plane, the equation

$$(3.1) \qquad z = tz_1 + (1-t)z_2, \qquad t \quad \text{real},$$

describes all the points on the straight line joining z_1 and z_2. The point z divides the line segment from z_1 to z_2 in the ratio

$$(3.2) \qquad r = (1-t)/t.$$

The distance from z_1 to z_2 is $\rho = |z_1 - z_2|$ or

$$(3.3) \qquad \rho = \sqrt{(z_1 - z_2)(\bar{z}_1 - \bar{z}_2)}.$$

The determinant

$$(3.4) \qquad D = \begin{vmatrix} z & \bar{z} & 1 \\ z_1 & \bar{z}_1 & 1 \\ z_2 & \bar{z}_2 & 1 \end{vmatrix}$$

vanishes for $z = z_1$ and for $z = z_2$ and since D is linear in x and y, the equation

$$(3.5) \qquad D = 0$$

must be the equation in conjugate coordinates of the

7

straight line through the points z_1 and z_2. Upon expansion, (3.5) can be written in the following forms:

$$(3.6) \quad z(\bar{z}_1 - \bar{z}_2) - \bar{z}(z_1 - z_2) + z_1\bar{z}_2 - z_2\bar{z}_1 = 0$$

or

$$(3.7) \quad \bar{z} = \left(\frac{\bar{z}_1 - \bar{z}_2}{z_1 - z_2}\right)(z - z_2) + \bar{z}_2; \qquad \frac{\bar{z} - \bar{z}_2}{z - z_2} = \frac{\bar{z}_1 - \bar{z}_2}{z_1 - z_2};$$

$$(3.8) \quad \bar{z} = \left(\frac{\bar{z}_1 - \bar{z}_2}{z_1 - z_2}\right)z + \left(\frac{z_1\bar{z}_2 - z_2\bar{z}_1}{z_1 - z_2}\right).$$

We can therefore write

$$(3.9) \qquad\qquad \bar{z} = Az + B,$$

where

$$(3.10) \qquad A = \frac{\bar{z}_1 - \bar{z}_2}{z_1 - z_2}, \qquad B = \frac{z_1\bar{z}_2 - z_2\bar{z}_1}{z_1 - z_2},$$

as an equation for the line through z_1 and z_2. If we use polar coordinates and write $z_2 = z_1 + \rho e^{i\theta}$ (see Fig. 3.1), then $\bar{z}_2 - \bar{z}_1 = \rho e^{-i\theta}$ so that

$$(3.11) \qquad A = e^{-2i\theta}, \qquad |A| = 1.$$

If we consider the line segment from z_2 to z_1, and if we write

$$z_1 = z_2 + \rho e^{i(\theta+\pi)}$$

$$= z_2 - \rho e^{i\theta},$$

then we arrive at (3.11) once again. Accordingly, the quantity A gives the "orientation" of the *undirected* line joining z_1 and z_2 and can serve as the complex analogue of

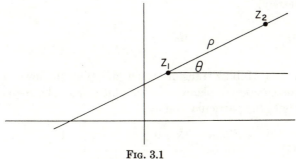

FIG. 3.1

the slope. The quantity A will be called the *clinant* of the line. Note that (3.9) yields

$$(3.12) \qquad \frac{d\bar{z}}{dz} = A,$$

which is a direct analog of the real situation

$$(3.13) \quad dy/dx = \lambda = \text{slope of the line} \quad y = \lambda x + b.$$

From (3.7), the equation of the straight line which passes through z_0 and has clinant A is given by

$$(3.14) \qquad \bar{z} = A(z - z_0) + \overline{z_0}.$$

The relationship between the clinant A and the slope λ of a straight line is as follows:

$$(3.15) \qquad A = \frac{\bar{z}_2 - \bar{z}_1}{z_2 - z_1} = \frac{(x_2 - x_1) - i(y_2 - y_1)}{(x_2 - x_1) + i(y_2 - y_1)}$$

$$= \frac{1 - i\lambda}{1 + i\lambda} = \frac{1 - i\tan\theta}{1 + i\tan\theta} = e^{-2i\theta}.$$

Inversely,

(3.16) $$\lambda = -i\,\frac{1-A}{1+A}.$$

This is a Möbius transformation, mapping the lower half of the complex λ-plane onto the unit disc of the A-plane. We note the particular values:

(3.17)

Slope	λ	0	1	∞	-1
Clinant	A	1	$-i$	-1	i

The angle ψ from a line with slope λ_1 and clinant A_1 to a line with slope λ_2 and clinant A_2 is given by

(3.18) $$\tan\psi = \frac{\lambda_2 - \lambda_1}{1 + \lambda_1\lambda_2} = i\,\frac{A_1 - A_2}{A_1 + A_2}.$$

It follows from this that two lines are parallel if and only if

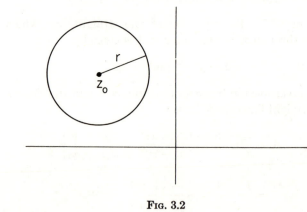

Fig. 3.2

their clinants are equal:

$$(3.19) \qquad A_1 = A_2,$$

and two lines are perpendicular if and only if their clinants are negatives:

$$(3.20) \qquad A_1 = -A_2.$$

Since for any three numbers z_1, z_2, z_3,

$$(3.21) \qquad \Delta = \begin{pmatrix} z_1 & \bar{z}_1 & 1 \\ z_2 & \bar{z}_2 & 1 \\ z_3 & \bar{z}_3 & 1 \end{pmatrix}$$

$$= \begin{pmatrix} x_1 & y_1 & 1 \\ x_2 & y_2 & 1 \\ x_3 & y_3 & 1 \end{pmatrix} \begin{pmatrix} 1 & 1 & 0 \\ i & -i & 0 \\ 0 & 0 & 1 \end{pmatrix},$$

by taking determinants we have

$$(3.22) \qquad |\Delta| = -4i \text{ Area } (z_1, z_2, z_3),$$

where Area (z_1, z_2, z_3) is the *signed area* of the triangle whose vertices are at z_1, z_2, z_3 in that order.

The circle. The equation of the circle with center at z_0 (see Fig. 3.2) and radius r can be written as $|z - z_0|^2 = r^2$ or $(z - z_0)(\bar{z} - \bar{z}_0) = r^2$. Hence

$$(3.23) \qquad \bar{z} = \bar{z}_0 + \frac{r^2}{z - z_0}$$

is its equation in conjugate coordinates.

Straight lines can be regarded as the "limits" of certain families of circles. Consider, for example, the family of circles of radius $s > 0$ and center at $z = is$. In conjugate coordinates,

$$\bar{z} = \frac{z}{1 + (iz/s)}.$$

As $s \to \infty$, the limiting form is $\bar{z} = z$, the equation of the x axis.

CHAPTER 4

THE NINE-POINT CIRCLE

Given a triangle T, the three midpoints of the three sides, the three feet of the altitudes from each vertex to the opposite side, and the three midpoints between the orthocenter (i.e., the intersection of the altitudes) and the vertices all lie on a circle. The radius of this circle is half that of the circle circumscribing T. The center of the nine-point circle lies half way between the circumcenter and the orthocenter. This is the *theorem of the nine-point circle* (see Fig. 4.1).

In proving this theorem it is advantageous to select the circumcircle as the unit circle $z\bar{z} = 1$. Let, therefore, the vertices of T be z_1, z_2, z_3 and note that $\bar{z}_i = z_i^{-1}$. For convenience, set

$$s = z_1 + z_2 + z_3$$

(4.1) $$t = z_1z_2 + z_2z_3 + z_3z_1$$

$$p = z_1z_2z_3.$$

The center of gravity of T (the intersection of the medians) is at

(4.2) $$z = \tfrac{1}{3}(z_1 + z_2 + z_3) = \tfrac{1}{3}s.$$

13

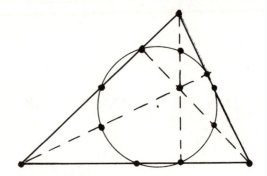

Fig. 4.1—The nine-point circle

The clinant of the side $[z_2 z_3]$ is

$$\frac{\bar{z}_2 - \bar{z}_3}{z_2 - z_3} = \frac{1/z_2 - 1/z_3}{z_2 - z_3} = -\frac{1}{z_2 z_3}.$$

Hence, the equation of this side is by (3.7)

$$\bar{z} = -\frac{1}{z_2 z_3}(z - z_2) + \bar{z}_2$$

or

(4.3) $$\bar{z} = -\frac{1}{z_2 z_3}(z - z_2) + \frac{1}{z_2}.$$

The perpendicular to this side through the vertex z_1 has the clinant $1/(z_2 z_3)$ and its equation is

(4.4) $$\bar{z} = \frac{1}{z_2 z_3}(z - z_1) + \bar{z}_1 \quad \text{or} \quad \bar{z} = \frac{1}{z_2 z_3}(z - z_1) + \frac{1}{z_1}.$$

Solving (4.3) and (4.4) simultaneously, we find that the

foot of the altitude is located at the point

$$(4.5) \quad z = \tfrac{1}{2}\left\{z_1 + z_2 + z_3 - \frac{z_2 z_3}{z_1}\right\} = \tfrac{1}{2}\left(s - \frac{p}{z_1^2}\right).$$

Two of the altitudes have equations

$$\bar{z} = \frac{1}{z_2 z_3}\,(z - z_1) + \frac{1}{z_1}\,, \qquad \bar{z} = \frac{1}{z_3 z_1}\,(z - z_2) + \frac{1}{z_2}\,.$$

Solving simultaneously we obtain their intersection

$$(4.6) \qquad\qquad z = z_1 + z_2 + z_3 = s.$$

This is the location of the orthocenter.

Note that the circumcenter is $z = 0$, the center of gravity (intersection of the medians) is $z = s/3$, and the orthocenter is $z = s$. Therefore the center of gravity lies $\tfrac{1}{3}$ of the way from the circumcenter to the orthocenter. The line joining these three points is called the *Euler line* of T. Its clinant is $(\bar{s} - \bar{0})/(s - \bar{0}) = \bar{s}/s$ and its equation is

$$(4.7) \qquad\qquad \bar{z} = \frac{\bar{s}}{s}\,z,$$

which can be written in the neat form

$$(4.8) \qquad\qquad \text{Im}\,(s\bar{z}) = 0.$$

(The real number

$$\text{Im}(s\bar{z}) = \begin{vmatrix} \text{Re} & z & \text{Im} & z \\ \text{Re} & s & \text{Im} & s \end{vmatrix}$$

is known as the *exterior product* of the complex numbers z and s and is designated by $z * s$.)

The nine points of the "nine-point circle" are therefore

$$(4.9) \quad \tfrac{1}{2}(z_1 + z_2), \qquad \tfrac{1}{2}(z_2 + z_3), \qquad \tfrac{1}{2}(z_3 + z_1)$$

$$\tfrac{1}{2}\left(s - \frac{p}{z_1^2}\right), \qquad \tfrac{1}{2}\left(s - \frac{p}{z_2^2}\right), \qquad \tfrac{1}{2}\left(s - \frac{p}{z_3^2}\right)$$

$$\tfrac{1}{2}(s + z_1), \qquad \tfrac{1}{2}(s + z_2), \qquad \tfrac{1}{2}(s + z_3).$$

Consider next the circle

$$(4.10) \qquad \bar{z} = \frac{1/4}{z - (s/2)} + \frac{\bar{s}}{2}.$$

Its radius is one-half that of the circumscribing circle and its center is at $z = s/2$. The center therefore lies on the Euler line halfway between the circum- and orthocenters. It is now a triviality to verify that the nine points in (4.9) satisfy (4.10).

Let z_1 be on the unit circle. The clinant of Oz_1 is $\bar{z}_1/z_1 = 1/z_1^2$. The equation of the tangent to the circle at z_1 is, by (3.7),

$$\bar{z} = -\frac{1}{z_1^2}(z - z_1) + \bar{z}_1 = -\frac{z}{z_1^2} + \frac{2}{z_1}.$$

Solving this equation with the equation of the tangent at z_2 we arrive at: $z = (2z_1z_2)/(z_1 + z_2)$. Call the triangle Δ formed by the 3 tangents to the unit circle at z_1, z_2, z_3 a *tangential triangle*. Thus, the coordinates of the vertices of the tangential triangle Δ are

$$\frac{2z_1z_2}{z_1 + z_2}, \qquad \frac{2z_2z_3}{z_2 + z_3}, \qquad \frac{2z_3z_1}{z_3 + z_1}.$$

The midpoints $z_i{}^*$ of the sides of Δ are computed to be

(after some reduction)

$$(4.11) \quad z_i{}^* = \frac{t^2}{st - p} - \frac{p^2}{(st - p)z_i{}^2} \qquad i = 1, 2, 3.$$

Note that $\bar{s} = t/p, \bar{t} = s/p, p\bar{p} = 1$. Furthermore $\overline{st - p} = (\bar{z}_1 + \bar{z}_2)(\bar{z}_2 + \bar{z}_3)(\bar{z}_3 + \bar{z}_1) = (st - p)/p^2$. Hence, $\overline{t^2/(st - p)} = s^2/(st - p)$. The equation, in conjugate coordinates, of the circle passing through $z_i{}^*$ (the nine-point circle of Δ) is

$$(4.12) \quad \left(z - \frac{t^2}{st - p}\right)\left(\bar{z} - \frac{s^2}{st - p}\right) = \frac{p^2}{(st - p)^2}$$

$$= \frac{1}{(st - p)(\overline{st - p})},$$

as is verified by substituting (4.11) in (4.12).

Solving (4.12) simultaneously with $z\bar{z} = 1$, we obtain

$$(4.13) \qquad s^2z^2 - 2stz + t^2 = 0.$$

If we assume that the triangle z_1, z_2, z_3 is not equilateral, then $s \neq 0$. Since the discriminant of (4.13) is $4s^2t^2 - 4s^2t^2 = 0$, it follows that (4.13) has a double root at $z = t/s$. This implies that the nine-point circle of Δ is tangent to the unit circle at this point. If $s = 0$, the two circles coincide.

We can now wrap up this discussion in the following way. Given any triangle, there are four circles that are simultaneously tangent to the three sides. One circle is inside the triangle; this is the *inscribed circle*. Three are exterior to the triangle; these are the *escribed circles*. The nine-point circle of any (nonequilateral) triangle is tangent to the inscribed and to the escribed circles. This is *Feuerbach's*

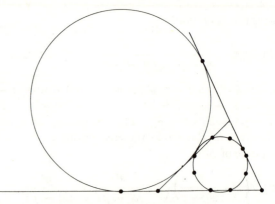

FIG. 4.2—Feuerbach's Theorem

Theorem (see Fig. 4.2). For the equilateral triangle, the nine-point circle, of course, coincides with the inscribed circle and is tangent to all the escribed circles.

If one adds to the nine points already distinguished, the four points of tangency just described, one might speak of the "thirteen-point circle." Some authorities say there are 31 and some say there are 43 points of significance on the nine-point circle. One can go on in this vein multiplying miracles and proving many other theorems of triangle and circle geometry. However, we have other fish to fry.

CHAPTER 5

THE SCHWARZ FUNCTION FOR AN ANALYTIC ARC

From straight lines and circles, we turn to analytic arcs and curves. We suppose that the arc C is written in rectangular form

$$(5.1) \qquad f(x, y) = 0.$$

In conjugate coordinates, this becomes

$$(5.2) \qquad f\left(\frac{z + \bar{z}}{2}, \frac{z - \bar{z}}{2i}\right) \equiv g(z, \bar{z}) = 0.$$

Example. Consider the real conic written in the matrix form

$$(5.3) \qquad X'AX = 1,$$

where

$$X = \begin{pmatrix} x \\ y \end{pmatrix}, \qquad A = \begin{pmatrix} a & b \\ b & c \end{pmatrix},$$

and the prime designates the transpose. If $Z = \begin{pmatrix} z \\ \bar{z} \end{pmatrix}$, then $Z = MX$ with

$$M = \begin{pmatrix} 1 & i \\ 1 & -i \end{pmatrix}.$$

From (2.4), $X = \frac{1}{2}M^*Z$, $X' = \frac{1}{2}Z'\bar{M}$ so that,

$$(5.4) \qquad Z'(\bar{M}AM^*)Z = 4$$

is the equation of the conic in conjugate coordinates.

If

$$(5.5) \qquad \delta = \det(\bar{M}AM^*)$$

then

$$(5.6) \qquad \delta = -4 \det A.$$

Since the conic is an ellipse or a hyperbola according as the discriminant $\delta > 0$ or $\delta < 0$, the transformation to conjugate coordinates switches the "formal type" of the quadratic form. As we shall see later, this switch also occurs when we consider differential operators of 2nd order and is of considerable importance in this theory.

Suppose, for the moment, that the function $g(z, \bar{z})$ of (5.2) is an irreducible polynomial of a certain degree (i.e., g cannot be expressed in the form $g \equiv g_1 g_2$ where g_1 and g_2 are real polynomials and whose degree ≥ 1). Write

$$(5.7) \qquad g(u, v) = \sum_{j,k=0}^{m,n} a_{jk} u^j v^k$$

and

$$(5.8) \qquad \bar{g}(u, v) = \sum_{j,k=0}^{m,n} \bar{a}_{jk} u^j v^k.$$

Along C we have $g(z, \bar{z}) = 0$, and by conjugating this equation we have $\overline{g(z, \bar{z})} \equiv \bar{g}(\bar{z}, z) = 0$. It follows by a well-known theorem of algebra that the two polynomials $g(z, \bar{z})$ and $\bar{g}(\bar{z}, z)$ must be proportional (see, e.g., Bôcher, p. 211). Thus,

$$(5.9) \qquad \bar{g}(\bar{z}, z) = \lambda g(z, \bar{z}), \qquad \lambda = \text{const.} \neq 0.$$

Polynomials with the property (5.9) are called *self-conjugate*.

Example. For the central conic (5.4), we have $g(z, \bar{z}) = (a - c - 2bi)z^2 + 2(a + c)z\bar{z} + (a - c + 2bi)\bar{z}^2 - 4$ with a, b, c real. This satisfies (5.9).

Self-conjugacy is therefore a necessary condition that a polynomial in conjugate coordinates z, $\bar{z}$ represent a real curve. It is not sufficient, as is seen from the example $g(z, \bar{z}) = z\bar{z} + 1$.

We now drop the polynomial hypothesis and assume only that g is an analytic function of z and $\bar{z}$. If at z_0, a point of the curve C, we have $\partial g/\partial \bar{z} \mid_{z_0} \neq 0$, then by the implicit function theorem, we may solve uniquely for $\bar{z}$ in terms of z to obtain

$$(5.10) \qquad \bar{z} = S(z),$$

where $S(z)$ is a regular analytic function of z in some neighborhood of z_0: $\mid z - z_0 \mid \leqq \rho(z_0)$. (See Fig. 5.1.) At z_0, we have $\bar{z}_0 = S(z_0)$ and the identity (5.10) persists along C in a neighborhood of z_0.

The condition $\partial g/\partial \bar{z} \neq 0$ is equivalent to

$$\frac{\partial}{\partial \bar{z}} f\left(\frac{z + \bar{z}}{2}, \frac{z - \bar{z}}{2i} \right) = \frac{1}{2}\left(\frac{\partial f}{\partial x} + i\frac{\partial f}{\partial y} \right) \neq 0$$

or to asserting that $\partial f/\partial x = 0$, $\partial f/\partial y = 0$ do not occur simultaneously. A point of C at which $\partial g/\partial \bar{z} \neq 0$ is called a *regular point of C*.

If, along a portion of a simple arc, endpoints included, $\partial g/\partial \bar{z} \neq 0$, we may solve for $\bar{z}$ at each point and obtain a regular $S(z)$. Thus, by a standard argument of analytic continuation, $S(z)$ can be defined as the branch of a single-valued analytic function in a strip-like region con-

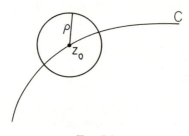

Fig. 5.1

taining the arc in its interior. (See Fig. 5.2.) If C intersects itself, then the striplike region will intersect itself; that is, $S(z)$ can be defined in a strip-like region lying on a Riemann surface. (See Fig. 5.3.)

If C is a simple closed analytic curve, then $S(z)$ is a branch of a single-valued analytic function in an annulus-like region containing C in its interior. (See Fig. 5.4.) It is possible that $S(z)$ may be continuable analytically to other portions of the plane as a single- or a multivalued function. For example, if g is a polynomial, then $S(z)$ is generally a multivalued algebraic function of z.

If $z_0 = 0$ lies on C then $g(0, 0) = 0$, and $S(z)$ has the

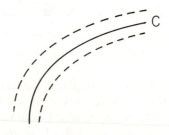

Fig. 5.2

FIG. 5.3

following representation as a line integral:

$$(5.10') \qquad S(z) = \frac{1}{2\pi i} \int_{\Gamma} \frac{w g_w(z, w)}{g(z, w)} \, dw,$$

where Γ is a suitably chosen contour surrounding $w = 0$. (See Notes, p. 211.)

We shall call $S(z)$ the Schwarz Function of C, for reasons which will emerge subsequently. By the uniqueness

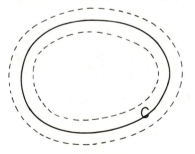

FIG. 5.4

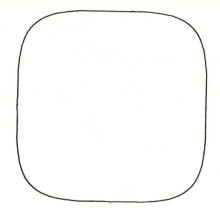

Fig. 5.5

theorem for analytic functions, an analytic function is determined uniquely by the values it takes along an arc. The Schwarz Function for C can therefore be defined alternatively as *the unique analytic function $S(z)$ which at each point z along C takes on the value $\bar{z}$.*

We shall next give a list of the Schwarz Functions of a number of familiar curves. Our list includes only curves whose Schwarz Functions are expressible in "elementary" terms. It is intended merely to be suggestive and not exhaustive.

The *straight line* passing through z_1 and z_2

$$(5.11) \qquad \bar{z} = S(z) = \frac{\bar{z}_1 - \bar{z}_2}{z_1 - z_2} z + \frac{z_1 \bar{z}_2 - z_2 \bar{z}_1}{z_1 - z_2}.$$

The *circle* of radius r, center at z_0

$$(5.12) \qquad \bar{z} = S(z) = \frac{r^2}{z - z_0} + \bar{z}_0.$$

The *ellipse* $(x^2/a^2) + (y^2/b^2) = 1$, $(a > b)$

$$(5.13) \quad \bar{z} = S(z) = \frac{a^2 + b^2}{a^2 - b^2} z + \frac{2ab}{b^2 - a^2} \sqrt{z^2 + b^2 - a^2}.$$

The *rectangular hyperbolas* $x^2 - y^2 = a^2$ with common asymptotes $y = \pm x$ and eccentricity $e = \sqrt{2}$ have the Schwarz Function

$$(5.13') \quad \bar{z} = S(z) = \sqrt{2a^2 - z^2} = \sqrt{\alpha^2 - z^2}, \quad \alpha = ea.$$

The *general conic section*: see (5.4).
The "L^4 *Gauge Curve*" $x^4 + y^4 = 1$ (see Fig. 5.5).

$$(5.14) \quad \bar{z} = S(z) = (-3z^2 + 2^{3/2}(z^4 + 1)^{1/2})^{1/2}.$$

(The curves $(x/a)^m + (y/b)^m = 1$ are sometimes called the *Lamé curves*.)

In preparation for the next several curves, observe the identity

$$\cos n\theta = \tfrac{1}{2}(e^{in\theta} + e^{-in\theta}) = \frac{1}{2r^n} \left[(re^{i\theta})^n + (re^{-i\theta})^n \right]$$

$$= \frac{z^n + \bar{z}^n}{2r^n}.$$

Hence, if n is even, $\cos n\theta = (z^n + \bar{z}^n)/2(z\bar{z})^{n/2}$.

The *Rose* R_{2m}: $r^{2m} = a + b \cos 2m\theta$, $0 < |b| < a$, $m = 1, 2, \cdots$.
Using the above identity we obtain

$$z^m \bar{z}^m = a + b \left(\frac{z^{2m} + \bar{z}^{2m}}{2z^m \bar{z}^m} \right).$$

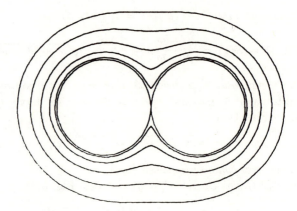

FIG. 5.6

Hence,

$$(5.15) \quad \bar{z} = S(z) = z \left[\frac{a + \sqrt{a^2 - b^2 + 2bz^{2m}}}{2z^{2m} - b} \right]^{1/m}.$$

Special instances of this are $m = 1$: the rose R_2, also called the *bicircular quartic* (see Fig. 5.6), which we can write in the form $r^2 = a^2 + 4\epsilon^2 \cos^2 \theta$.

$$(5.16) \quad \bar{z} = S(z) = \frac{z(a^2 + 2\epsilon^2) + z\sqrt{a^4 + 4a^2\epsilon^2 + 4\epsilon^2 z^2}}{2(z^2 - \epsilon^2)}.$$

Selection of $m = 2$ yields the rose R_4, which we write in the form $r^4 = a^4 + 2b^4 \cos 4\theta$ $(a^4 > 2b^4)$.

In this case,

$$(5.17) \quad \bar{z}^2 = S^2(z) = \frac{a^4 z^2 + z^2 \sqrt{4b^4 z^4 + a^8 - 4b^8}}{2(z^4 - b^4)}.$$

If $p_n(z) = (z - z_1)(z - z_2) \cdots (z - z_n)$, the curves

$| p_n(z) | = r^n$ are *generalized lemniscates* $\mathcal{L}$. If we set $\bar{p}_n(z) = (z - \bar{z}_1) \cdots (z - \bar{z}_n)$, then $p_n(z)\bar{p}_n(\bar{z}) = r^{2n}$ so that

$$\bar{p}_n(S(z)) = \frac{r^{2n}}{p_n(z)} .$$

In particular, if $p_n(z) = z^n - 1$, we have

$$(5.18) \qquad \bar{z} = S(z) = \sqrt[n]{\frac{z^n + r^{2n} - 1}{z^n - 1}} .$$

For $n = 2$, we have the *Ovals of Cassini*.

We shall exhibit several additional analytic curves with "elementary" transcendental Schwarz Functions in the course of the book.

CHAPTER 6

GEOMETRICAL INTERPRETATION
OF THE SCHWARZ FUNCTION;
SCHWARZIAN REFLECTION

Reflection in a straight line. Let z_1 and z_2 be two distinct points. The transformation

$$T(z) = \frac{|z_1 - z_2|}{(z_1 - z_2)}(z - z_2)$$

is a rigid motion which brings z_2 to the origin and z_1 to the x-axis. Its inverse is $T^{-1}(z) = ((z_1 - z_2)/|z_1 - z_2|)z + z_2$. The transformation $R(z) = \bar{z}$ is a reflection in the x axis. Hence the composite function $T^{-1}RT$ reflects z in the line l determined by z_1 and z_2. It is easily computed to be

$$(6.1) \qquad z^* = T^{-1}RT(z) = \frac{z_1 - z_2}{\bar{z}_1 - \bar{z}_2}(\bar{z} - \bar{z}_2) + z_2.$$

Comparing this with (5.11) or (3.7) we see that

$$(6.1') \qquad\qquad z^* = \overline{S(z)},$$

where $S(z)$ is the Schwarz Function of l. (See Fig. 6.1.)

Reflection or inversion in a circle $C: |z - z_0| = r$. The definition is as follows: Let $z \neq z_0$. Draw the half ray l^+ from z_0 through z. Locate the point z^* on l^+ such that

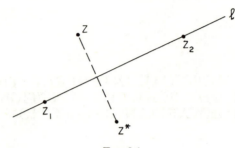

FIG. 6.1

$|z^* - z_0| \, |z - z_0| = r^2$. We must therefore have $(z^* - z_0) = \sigma(z - z_0)$, σ positive. Hence $\sigma \, |z - z_0|^2 = r^2$, so that

$$(6.2) \qquad z^* = z_0 + \frac{r^2(z - z_0)}{|z - z_0|^2} = z_0 + \frac{r^2}{\bar{z} - \bar{z}_0}.$$

Comparing this with (5.12), we see again that

$$(6.2') \qquad\qquad z^* = \overline{S(z)},$$

where $S(z)$ is the Schwarz Function of the circle. (See Fig. 6.2.)

Reflection in a general analytic arc. We now seek a similar geometric interpretation of the Schwarz Function for a general analytic arc. To do so, we shall have to be more precise as to what constitutes an *analytic arc*.

Given an arc C expressed in terms of a real parameter t in the form

$$(6.3) \qquad \begin{cases} x = f_1(t) \\ y = f_2(t) \end{cases} \qquad 0 \leqq t \leqq 1,$$

set

$$(6.4) \qquad z(t) = x + iy = f_1(t) + if_2(t) = f(t).$$

Then C is said to be *a simple analytic arc* if (a) $z(t_1) = z(t_2)$ only when $t_1 = t_2$, (b) $f_1(t)$ and $f_2(t)$ are real analytic functions of t for $0 \leq t \leq 1$, and (c) $z'(t) = f_1'(t) + if_2'(t) \neq 0, 0 \leq t \leq 1$.

Now for any $t_0, 0 \leq t_0 \leq 1, f(t)$, considered as a function of the complex variable t, is analytic in some circle $|t - t_0| \leq \lambda(t_0)$. Since $f'(t_0) \neq 0$, $f(t)$ maps some subcircle $|t - t_0| \leq \lambda_1 \leq \lambda$ *one to one* conformally onto a region R_{z_0} containing the point $z_0 = f(t_0)$. Any point $z \in R_{z_0}$ therefore is the image of a unique point t in the t-plane: $z = f(t) = f_1(t) + if_2(t)$. Suppose now that z^* is the image of $\bar{t}$ under f:

$$(6.5) \qquad z^* = f(\bar{t}) = f_1(\bar{t}) + if_2(\bar{t});$$

then z^* is called the *Schwarz reflection of z* in the analytic

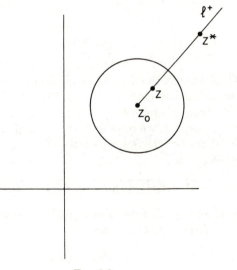

FIG. 6.2

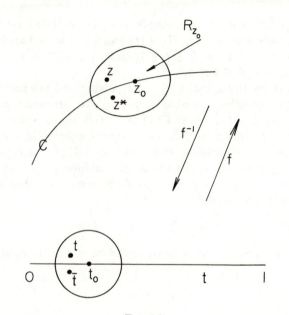

Fig. 6.3

arc C. Note that the Schwarz reflection is defined only for points sufficiently close to C. (See Fig. 6.3.)

Reflection therefore is defined by the sequence $z \to t \to \bar{t} \to z^*$. If we start with z^*, we obtain $z^* \to \bar{t} \to \bar{\bar{t}} = t \to z$. Therefore the Schwarz reflection of z^* must be z.

Furthermore,

$$(6.6) \qquad \overline{z^*} = \overline{f_1(\bar{t}) + if_2(\bar{t})} = \overline{f_1(\bar{t})} - \overline{if_2(\bar{t})}.$$

Since f_1 and f_2 are real-valued analytic functions on the real line, $f_i(\bar{t}) = \overline{f_i(t)}$ and hence

$$(6.7) \qquad \overline{z^*} = f_1(t) - if_2(t).$$

Therefore,

$$(6.8) \quad x = f_1(t) = \frac{z + \bar{z}^*}{2}, \qquad y = f_2(t) = \frac{z - \bar{z}^*}{2i}.$$

Setting this in the rectangular equation of $C: F(x, y) = 0$, we obtain

$$F\left(\frac{z + \bar{z}^*}{2}, \frac{z - \bar{z}^*}{2i}\right) = 0 \quad \text{or} \quad g(z, \bar{z}^*) = 0.$$

But solving this for $\bar{z}^*$ (since $\partial g/\partial \bar{z} \neq 0$) yields

$$(6.9) \qquad\qquad \overline{z^*} = S(z),$$

so that

$$(6.10) \qquad\qquad z^* = \overline{S(z)}.$$

This tells us that *the conjugate of the Schwarz Function of an analytic arc yields the Schwarzian reflection of a point in that arc.*

Insofar as the Schwarz Function is defined uniquely by the arc, we learn that the operation of reflection is independent of the parameterization of the arc. Schwarzian reflection is an *anti-conformal* transformation, i.e., one that reverses the sense of angles.

As we have seen, the reflection of the reflection of a point z must be z itself. Hence it follows that

$$(6.11) \qquad\qquad \overline{S(\overline{S(z)})} \equiv z.$$

This is a functional equation satisfied by $S(z)$.

Any solution of (6.11) will be called an *involutory* (or *Hermitian involutory*) *function.* We shall write this equation in a slightly different form.

Suppose that $h(z)$ is an analytic function of a complex

variable defined in a region R of the complex plane. By $\bar{R}$ designate the region R reflected in the real axis: $\bar{R} = \{z: \bar{z} \in R\}$. Define a function $\bar{h}(z)$ in $\bar{R}$ by means of the equation

$$(6.12) \qquad \bar{h}(z) = \overline{h(\bar{z})}, \qquad z \in \bar{R}.$$

It will be very useful to develop a few properties of the *conjugate function* $\bar{h}$. The function $\bar{h}(z)$ is analytic in $\bar{R}$. It satisfies the equation

$$(6.13) \qquad \overline{h(z)} = \bar{h}(\bar{z}), \qquad z \in R.$$

If, locally, $h(z) = \sum_{k=0}^{\infty} a_k(z - z_0)^k$, then $\overline{h(z)} = \sum_{k=0}^{\infty} \bar{a}_k(\bar{z} - \bar{z}_0)^k$ so that

$$(6.14) \qquad \bar{h}(z) = \sum_{k=0}^{\infty} \bar{a}_k(z - \bar{z}_0)^k.$$

Note that if the a_k and z_0 are real, then of course $\bar{h} = h$. We have

$$(6.15) \qquad \overline{\bar{h}}(z) = h(z),$$

and furthermore,

$$(6.16) \quad (\bar{h})' = \overline{(h')}. \qquad \text{(The prime designates differentiation.)}$$

We have

$$(6.17) \qquad \overline{g(h(z))} = \bar{g}(\overline{h(z)}) = \bar{g}(\bar{h}(\bar{z})).$$

We also have $\overline{gh}(z) = \overline{g(h(\bar{z}))} = \bar{g}(\bar{h}(z))$, and therefore

$$(6.18) \qquad \overline{gh} = \bar{g}\bar{h}.$$

If g has an inverse so that $g^{-1}g = gg^{-1} = I$, where I is the *identity function*, then $\bar{g}^{-1}\bar{g} = \overline{g^{-1}g} = \bar{I} = I$ so that

$$(6.19) \qquad \overline{(g^{-1})} = (\bar{g})^{-1}.$$

Using this notation for conjugate functions, we have from (6.11)

$$\bar{S}(S(z)) \equiv z,$$

which we write as

$$(6.20) \qquad\qquad \bar{S}S = I.$$

From this it follows that

$$(6.21) \quad \bar{S} = S^{-1}, \quad S = (\bar{S})^{-1} = \overline{(S^{-1})}, \quad \text{and} \quad S\bar{S} = I.$$

In the remainder of this essay, the juxtaposition of functional symbols as in (6.20) will often designate *functional composition*.* Addition, multiplication, and functional composition together comprise what some authors have called *trioperational algebra*. The relevant formal rules are

$$(1) \qquad\qquad f(gh) = (fg)h$$

$$(2)^, \qquad\qquad (f+g)h = fh + gh$$

$$(3) \qquad\qquad (f \cdot g)h = fh \cdot gh$$

$$(4) \qquad\qquad f^{-1}f = ff^{-1} = I$$

$$(5) \qquad\qquad (fg)^{-1} = g^{-1}f^{-1}.$$

Note that (2) and (3) cannot be turned around to say something about $h(f+g)$ or $h(f \cdot g)$.

To these three operations we add *conjugation* whose

---

* It is hoped that the usage will be sufficiently clear and that confusions with ordinary multiplication will not arise. (Many authors use $f \circ g$ for functional composition, but this notation is thought to be unnecessarily heavy in the present setting.)

rules are

(6)
$$\bar{\bar{f}} = f$$

(7)
$$\overline{f + g} = \bar{f} + \bar{g}$$

(8)
$$\overline{f \cdot g} = \bar{f} \cdot \bar{g}$$

(9)
$$\overline{fg} = \bar{f}\bar{g}$$

(10)
$$(\bar{f})^{-1} = \overline{(f^{-1})}.$$

We also add *differentiation* whose rules are

(11)
$$\overline{(f)'} = (\bar{f}')$$

(12)
$$(f + g)' = f' + g'$$

(13)
$$(f \cdot g)' = f \cdot g' + f' \cdot g$$

(14)
$$(fg)' = f'g \cdot g'.$$

A simple instance of (6.21) is provided by the Schwarz Function of a circle. A matrix formulation of this is more interesting than direct substitution, and we therefore digress to discuss the composition of linear fractional transformations by means of matrices. We shall represent the rational function (also referred to as a linear, bilinear, linear fractional, or Möbius transformation) $w = f(z) = (Az + B)/(Cz + D)$ by the matrix $\begin{pmatrix} A & B \\ C & D \end{pmatrix}$. Write $f \sim \begin{pmatrix} A & B \\ C & D \end{pmatrix}$. We assume that $\begin{vmatrix} A & B \\ C & D \end{vmatrix} \neq 0$, otherwise the numerator and denominator are proportional and f is constant. Notice that the function $f(z)$ is equally well represented by any matrix of the form

$$\lambda \begin{pmatrix} A & B \\ C & D \end{pmatrix} = \begin{pmatrix} \lambda A & \lambda B \\ \lambda C & \lambda D \end{pmatrix}$$

for any $\lambda \neq 0$. If now

$$u = g(w) = \frac{\alpha w + \beta}{\gamma w + \delta},$$

then

$$u = gf(z) = \frac{\alpha \left(\dfrac{Az + B}{Cz + D} \right) + \beta}{\gamma \left(\dfrac{Az + B}{Cz + D} \right) + \delta}$$

$$= \frac{(\alpha A + \beta C)z + (\alpha B + \beta D)}{(\gamma A + \delta C)z + (\gamma B + \delta D)}.$$

Therefore,

$$(6.22) \qquad gf \sim \begin{pmatrix} \alpha & \beta \\ \gamma & \delta \end{pmatrix} \begin{pmatrix} A & B \\ C & D \end{pmatrix}.$$

This relationship can be expressed in a slightly different form. If we use the notation M_f to designate the matrix $\begin{pmatrix} A & B \\ C & D \end{pmatrix}$, then

$$(6.22') \qquad M_{gf} = \lambda M_g M_f, \qquad \lambda \neq 0.$$

This convenient device will be used in a number of places in what follows.

For the circle $C \colon |z - z_0| = r$, in view of (5.12),

$$(6.23) \qquad S(z) \sim \begin{pmatrix} \bar{z}_0 & r^2 - |z_0|^2 \\ 1 & -z_0 \end{pmatrix}$$

and

$$\bar{S}(z) \sim \begin{pmatrix} z_0 & r^2 - |z_0|^2 \\ 1 & -z_0 \end{pmatrix}.$$

Multiplying these two matrices in the order $S\bar{S}$, we obtain

$$S\bar{S}(z) \sim \begin{pmatrix} r^2 & 0 \\ 0 & r^2 \end{pmatrix} \sim \frac{r^2 z}{r^2} \equiv z \qquad (\text{or } S\bar{S} = I).$$

We note in passing that the unit circle $\bar{z} = 1/z$ is represented by the matrix $E = \begin{pmatrix} 0 & 1 \\ 1 & 0 \end{pmatrix}$, sometimes called the *counteridentity*. The x axis $\bar{z} = z$ is represented by the identity $I = \begin{pmatrix} 1 & 0 \\ 0 & 1 \end{pmatrix}$, and one has $E^2 = I$.

From (6.20), we conclude that an "arbitrary" analytic function $S(z)$ cannot serve as a Schwarz Function. Hence, for general analytic $S(z)$, the roots of $\bar{z} = S(z)$ cannot fill out an analytic arc.

Example. If $\bar{z} = z^n$, $n > 1$, then $|z| = |z|^n$ so that, excepting $z = 0$, $|z| = 1$. Whence, $z = \exp(2\pi ji/(n+1))$, $j = 0, 1, \ldots, n$.

Given $n + 1$ distinct points in the complex plane z_i, one can readily find a polynomial $p_n(z)$ of degree $\leq n$ such that $p_n(z_i) = \bar{z}_i$, where $i = 1, 2, \ldots, n + 1$. This polynomial will not in general be the Schwarz Function for an arc.

Even if an analytic function $S(z)$ satisfies (6.20), it may not be the Schwarz Function of a real arc.

Example. If $S(z) = (z - a)/(z - 1)$, a real, then $\bar{S}S = I$. However, $\bar{z} = (z - a)/(z - 1)$ is equivalent to $(x - 1)^2 + y^2 = 1 - a$ so there is no solution in the real Gauss plane for $a > 1$.

We shall see later on that a necessary and sufficient

condition that $S(z)$ be a Schwarz Function is that it have a representation of the form (8.1).

Insofar as $S(z)$ is an analytic function of z, we may study and interpret its derivatives, analytic continuation, definition on Riemann surfaces, singularities, etc. We may also perform residue calculus, look at its representations and functional equations, and seek applications. This program will be carried out in the remainder of this book.

CHAPTER 7

THE SCHWARZ FUNCTION AND DIFFERENTIAL GEOMETRY

In this chapter, we shall relate the derivatives of the Schwarz Function of an analytic arc C to the slope, curvature, etc., of the arc. Let the point $z = re^{i\theta}$ lie on C. Then

(7.1) $\qquad r^2 = z\bar{z} = zS(z) = |S(z)|^2$ along C.

Since $\bar{z}/z = e^{-2i\theta}$, we have

(7.2) $\qquad \theta = \frac{i}{2} \log\left(\frac{\bar{z}}{z}\right) = \frac{i}{2} \log\left(\frac{S(z)}{z}\right)$, along C.

Along C we have,

(7.3) $\qquad \dfrac{d\bar{z}}{dz} = \dfrac{dx - idy}{dx + idy} = \dfrac{1 - iy'}{1 + iy'}$, where $y' = \dfrac{dy}{dx}$.

However, along C, $\bar{z} = S(z)$, and since the derivative of an analytic function is independent of the direction in which increments are taken, we have

(7.4) $\qquad S'(z) = \dfrac{d\bar{z}}{dz} = \dfrac{1 - iy'}{1 + iy'}$.

Since $dx - idy = \overline{dx + idy}$, it follows from (7.3) and

41

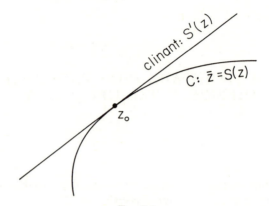

FIG. 7.1

(7.4) that

(7.5) $$| S'(z) | = 1 \quad \text{along } C.$$

The equality (7.5) may also be obtained as a consequence of the functional equation (6.20) (or (6.11)). Differentiating (6.20), we obtain $\bar{S}'(S(z))S'(z) \equiv 1$. Now along C, $S(z) = \bar{z}$, so that we get $\bar{S}'(\bar{z})S'(z) = 1$ or $\overline{S'(z)}\,S'(z) = 1$, which is (7.5).

Solving (7.4) yields

(7.6) $$y' = -i\,\frac{1 - S'(z)}{1 + S'(z)}.$$

Let the point z_0 lie on C. A line through z_0 with slope λ has the equation $\bar{z} = A(z - z_0) + \bar{z_0}$ where by (3.15), $A = (1 - i\lambda)/(1 + i\lambda)$. But if $\lambda = y'(x_0)$, then by (7.4), $A = S'(z_0)$. Hence, the equation of the tangent to C at z_0 is

(7.7) $$\bar{z} = S'(z_0) \cdot (z - z_0) + \bar{z_0}.$$

This is the analog of the point-slope formula. (See Fig. 7.1.) We may therefore speak of $S'(z_0)$ as the *clinant* of the arc C at z_0, and we see that the derivative of the Schwarz Function plays a role analogous to that of slope.

If ψ is the angle the tangent to C at z_0 makes with the real axis, then

$$\tan \psi = y' = -i\frac{1 - S'}{1 + S'}.$$

It follows from this that

(7.8) $\psi = \tfrac{1}{2} \arg S'(z_0).$

Thus, (7.8) and (7.5) completely identify $S'(z)$ along C.

If two analytic arcs with Schwarz Functions $S(z)$ and $T(z)$ meet at a common point z_0 (see Fig. 7.2), then the angle ψ from the first arc to the second is given by

$$\tan \psi = \frac{\lambda_2 - \lambda_1}{1 + \lambda_1\lambda_2},$$

where λ_1 and λ_2 are the respective slopes. Using (7.6) we

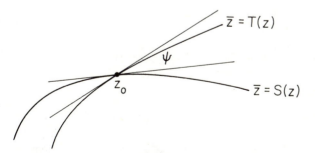

Fig. 7.2

find

$$(7.9) \qquad \tan \psi = i \left(\frac{S' - T'}{S' + T'} \right)_{z=z_0}$$

or

$$(7.10) \qquad \psi = \tfrac{1}{2}(\arg T' - \arg S').$$

We see that two arcs that have a common point are tangent there if and only if

$$(7.11) \qquad S' = T'$$

and are orthogonal if and only if

$$(7.12) \qquad S' = -T'.$$

Suppose that the arcs S, T, and U are concurrent at z_0 and the arc T bisects the angle from the arc S to the arc U. Then from (7.9),

$$\frac{S' - T'}{S' + T'} = \frac{T' - U'}{T' + U'}$$

or

$$(7.12a) \qquad (T')^2 = S'U' \quad \text{at } z_0$$

so that $T'(z_0)$ is the geometric mean of $S'(z_0)$ and $U'(z_0)$.
We have

$$(7.13) \quad ds^2 = dx^2 + dy^2 = (dx + idy)(dx - idy) = dz\overline{dz}$$
$$= dz d\bar{z} = dz S'(z) dz = S'(z)(dz)^2.$$

Hence

$$(7.13') \qquad dz/ds = 1/\sqrt{S'(z)}.$$

We next find interpretations for the 2nd derivative of the Schwarz Function. We have $d^2y/dx^2 = (dy'/dz) \cdot$

(dz/dx). From (7.6), $dy'/dz = 2iS''/(1 + S')^2$. Now, $dz/dx = (dx + idy)/dx = 1 + iy'$ so that by (7.6),

$$(7.14) \qquad \frac{dz}{dx} = \frac{2}{1 + S'} \cdot$$

Hence,

$$(7.15) \qquad \frac{d^2y}{dx^2} = \frac{4iS''}{(1 + S')^3} \cdot$$

If we designate the (signed) curvature of C by k, we have

$$(7.16) \quad k = \frac{d\psi}{ds} = \frac{d \tan^{-1} y'}{dx} \cdot \frac{dx}{dz} \cdot \frac{dz}{ds} = \frac{y''}{1 + y'^2} \cdot \frac{dx}{dz} \cdot \frac{dz}{ds} \cdot$$

Using (7.15), (7.14), $(7.13')$, we find that

$$(7.17) \qquad k = \frac{i}{2} \frac{S''}{(S')^{3/2}} \cdot$$

Hence, from (7.5),

$$(7.18) \qquad |k| = \tfrac{1}{2} |S''|, \qquad z \in C$$

and we may speak of S'' as being a "complex curvature" of C.

Example. A circle has constant curvature. Hence, differentiating (7.17), we find that $2S' \cdot S''' = 3(S'')^2$ is a differential equation satisfied by the Schwarz Function of any circle.

For a point z_0 on C we have

$$(7.19) \qquad S(z) = \sum_{n=0}^{\infty} \frac{S^{(n)}(z_0)}{n!} (z - z_0)^n,$$

and on the basis of previous relations, we can work out

the first few terms of (7.19). The first two are

$$(7.20) \quad S(z) \approx S(z_0) + S'(z_0)(z - z_0)$$

$$= \overline{z_0} + S'(z_0)(z - z_0)$$

$$= \text{the Schwarz Function of the tangent.}$$

The Schwarz Function of the normal is

$$(7.20') \qquad S_N(z) = \overline{z_0} - S'(z_0)(z - z_0).$$

Therefore in a neighborhood of z_0, the Schwarzian reflection of z in C is given approximately by

$$(7.21) \qquad z^* = \overline{S(z)} \approx \overline{S'(z_0)}(\bar{z} - \bar{z}_0) + z_0.$$

The expression on the right is the (ordinary) reflection of z in the tangent to C at z_0 (see (6.1)). Hence, *neglecting terms of the 2nd order or higher, Schwarzian reflection in an arc C coincides with ordinary reflection in the tangent to C.*

The higher derivatives of $S(z)$ may be expressed in terms of the higher differential invariants $k' = dk/ds$, $k'' = dk'/ds = d^2k/ds^2$, etc. We shall work out one more. We have

$$(7.22) \qquad k' = \frac{dk}{ds} = \frac{dk}{dz}\frac{dz}{ds},$$

whence from (7.17) and (7.13'),

$$(7.23) \qquad k' = \frac{i}{2}\frac{S'S''' - \frac{3}{2}(S'')^2}{(S')^3}.$$

In the theory of conformal mapping, the expression $w'''/(w')^2 - \frac{3}{2}(w'')^2/(w')^3$ is known as the *Schwarzian derivative* and is designated by $\{w, z\}$. It is a differential invariant for the group of all linear fractional transformations; i.e., if $W = (Aw + B)/(Cw + D)$, then $\{W, z\} = \{w, z\}$.

This can be shown by a formal computation. In particular, if $w(z) = z$, then $\{(Az + B)/(Cz + D), z\} = \{z, z\} = 0$. Furthermore, if $Z = (az + b)/(cz + d)$, then $\{w, z\} = (dZ/dz)^2\{w, Z\}$. In this notation, (7.23) becomes

$$(7.23') \qquad k' = \frac{i}{2}\{S, z\}.$$

Example. For a circle, $S(z)$ is a bilinear function. Hence $k' = \frac{1}{2}i\{S, z\} = 0$.

The series (7.19), which is the local expansion of $S(z)$, can be rewritten in terms of slopes, curvatures, etc. The first three terms are available through (7.4), (7.17), and (7.23). It simplifies matters somewhat to assume that C passes through $z_0 = 0$ and has slope 0 there. Therefore $S(0) = 0$, $S'(0) = 1$.

With this simplification,

$$(7.23'') \quad S(z) = z - ikz^2 + \left(-k^2 - \frac{i}{3}k'\right)z^3$$
$$+ \left[-\tfrac{5}{6}kk' - i(\tfrac{1}{12}k'' - k^3)\right]z^4$$
$$+ \left[(-\tfrac{1}{4}kk'' - \tfrac{1}{6}(k')^2 + k^4)\right.$$
$$\left. - i(\tfrac{1}{60}k''' - \tfrac{43}{40}k^2k')\right]z^5 + \cdots \mid z \mid < \rho.$$

Assuming again that $z_0 = 0$ and that the slope of C at z_0 is 0, consider the *circle of curvature* C_k of C at this point. It has radius $1/|k|$ and center at $z_1 = i/k$. Hence the Schwarz function of C_k is

$$(7.24) \quad S_k(z) = \frac{iz}{i - kz} = z - ikz^2 - k^2z^3 + \cdots.$$

Therefore,

$$(7.25) \qquad S_k(z) - S(z) = \frac{i}{3} k'z^3 + \cdots.$$

Hence, *the Schwarz Function for the circle of curvature coincides with the Schwarz Function of the arc up through terms of the 2nd order.* We have $S(z_0) = S_k(z_0)$, $S'(z_0) = S_k'(z_0)$, $S''(z_0) = S_k''(z_0)$ so that $S_k(z)$ is a rational approximant to $S(z)$ *in the sense of Padé.* Higher Padé approximants to $S(z)$ may be found, i.e., rational functions $R(z)$ for which $R^{(j)}(z_0) = S^{(j)}(z_0)$, $j = 0, 1, \cdots, q$. However, none of these will be the Schwarz Function for an arc. (See Chapter 10.)

CHAPTER 8

CONFORMAL MAPS, REFLECTIONS, AND THEIR ALGEBRA

In Chapter 6, we defined an analytic arc C as the image of the real segment $a \leqq t \leqq b$ under a one-to-one conformal map f. We found that if a point $z = f(t)$ is in a neighborhood of C, the reflection* of z in C is defined by $z^* = f(\bar{t})$. Since $S(z) = \overline{z^*}$, we have $S(z) = \overline{f(\bar{t})} = \bar{f}(t)$ and since $t = f^{-1}(z)$, we have $S(z) = \bar{f}(f^{-1}(z))$. We can write this as

$$(8.1) \qquad S = \bar{f}f^{-1},$$

from which we obtain

$$(8.1') \qquad S' = \bar{f}'f^{-1}/f'f^{-1}.$$

If the parameter t in $[0, 1]$ is changed by means of

$$(8.2) \qquad t = g(t'), \qquad 0 \leqq t' \leqq 1,$$

where g is real analytic on the real t' axis: $g = \bar{g}$, then we can compute $S(z)$ equally well on the basis of the com-

---

* The notion of reflection can be generalized to certain non-analytic arcs. It suffices to deal with mapping functions f for which $z^* = f f^{-1}(z)$ is defined. Such a generalization is relevant, e.g., to the theory of quasi-analytic functions and has been worked out in considerable detail. We shall not be able to expound these matters here.

posite mapping fg. From (8.1) this yields

$$(8.3) \quad S = \overline{fg}(fg)^{-1} = \bar{f}\bar{g}g^{-1}f^{-1} = \bar{f}gg^{-1}f^{-1} = \bar{f}f^{-1}.$$

Thus, as required, S is independent of the parameterization of C.

In a neighborhood of $t = 0$, we write $z = f(t) = at + bt^2 + ct^3 + \cdots$, $a \neq 0$. Inverting this series formally, we find $t = f^{-1}(z) = Az + Bz^2 + Cz^3 + \cdots$, where

$$aA = 1$$

$$a^3B = -b$$

$$a^5C = 2b^2 - ac$$

$$a^7D = 5abc - a^2d - 5b^3$$

$$a^9E = 6a^2bd + 3a^2c^2 + 14b^4 - a^3e - 21ab^2c$$

. . . .

Substituting,

$$S(z) = \bar{f}(f^{-1}(z))$$

$$(8.3a) \qquad = \frac{\bar{a}}{a}z + \left(\frac{a\bar{b} - \bar{a}b}{a^3}\right)z^2$$

$$+ \left(\frac{a(a\bar{c} - \bar{a}c) + 2b(\bar{a}b - a\bar{b})}{a^5}\right)z^3 + \cdots.$$

If f takes real values on the real axis then $a = \bar{a}$, $b = \bar{b}$, $\cdots$, and all the coefficients after the first vanish, for the t axis is mapped onto the x axis and $S(z) = z$.

If we write $S(z) = b_1z + b_2z^2 + b_3z^3 + \cdots$, then from

$\bar{S} = S^{-1}$ and from the coefficients of S^{-1}, we obtain

$$\bar{b}_1 = \frac{1}{b_1}$$

$$\bar{b}_2 = -\frac{b_2}{b_1{}^3}$$

(8.3b) $$\bar{b}_3 = \frac{2b_2{}^2 - b_1 b_3}{b_1{}^5}$$

$$\bar{b}_4 = \frac{5b_1 b_2 b_3 - b_1{}^2 b_4 - 5b_2{}^3}{b_1{}^7}$$

. . . .

Suppose now that $b_1 = 1$ and the coefficients $b_2, b_3, b_4, \cdots$ are all real, i.e., Im $b_2 =$ Im $b_3 = \cdots = 0$. Now from the above equations, $\bar{b}_2 + b_2 = 2 \text{ Re } b_2 = 0$. Hence $b_2 = 0$; $\bar{b}_3 + b_3 = 2 \text{ Re } b_3 = 2b_2{}^2 = 0$. Hence $b_3 = 0$. For general $n \geqq 2$, with $b_1 = 1$, $\bar{b}_n + b_n = 2 \text{ Re } b_n =$ a homogeneous polynomial in $b_2, b_3, \cdots, b_{n-1}$. Hence $b_n = 0$, $n = 2, 3, \cdots$ so that $S(z) = z$.

The Schwarz Function has an interpretation as a conformal map. We have $S(z) = \bar{f}(f^{-1}(z)) = \overline{f(\overline{f^{-1}(z)})}$. Now suppose that $a < t_0 < b$ and that $f(t)$ maps the disc $\mathcal{C}: |t - t_0| < \epsilon$ onto a region $\mathcal{G}$ in the z plane that contains $z_0 = f(t_0)$ and a portion of C in its interior. This will be the case with ϵ small enough. Then $S(z)$ maps $\mathcal{G}$ conformally onto $\bar{\mathcal{G}}$. For with $z \in \mathcal{G}$, $f^{-1}(z) \in \mathcal{C}$, $\overline{f^{-1}(z)} \in \mathcal{C}$, $f(\overline{f^{-1}(z)}) \in \mathcal{G}$, and finally, $\overline{f(\overline{f^{-1}(z)})} \in \bar{\mathcal{G}}$. All the mappings are one-to-one onto. Since $S(z)$ is schlicht (univalent) in $\mathcal{G}$, a consequence is that $S'(z) \neq 0$ throughout $\mathcal{G}$.

Instead of the circle $\mathcal{C}$ we can work with any simply connected region in the t-plane that is symmetric with respect to the real axis and lies within the region of univalence of f. This leads to the following notion which is of some use in what follows. The region $\mathcal{G}$ is called *conformally symmetric* with respect to the analytic arc C if its image under f^{-1} is symmetric with respect to the real axis.

It is often convenient, particularly when dealing with a simple closed analytic curve C, to obtain its Schwarz Function from the function which maps C onto the unit circle. Let C be such a closed curve and let $w = M(z)$ be any function performing a 1-1 conformal map of the interior of C in the z-plane onto the interior of the unit circle $|w| = 1$. Let $z = m(w)$ be the inverse mapping function.

By a well-known theorem in conformal mapping (see, e.g., Ahlfors [A1] p. 225), since C is an analytic curve, $M(z)$ can be continued across the boundary of C from the interior so as to be analytic and schlicht in a larger region with analytic boundary C'. Similarly, $m(w)$, and hence $\bar{m}(w)$, can be continued analytically into $|w| \leqq r$ for some $r > 1$. For z interior to C', $w = M(z)$ takes values in $|w| < r', r' > 1$. Hence $1/M(z)$ takes values in $|w| > 1/r'$. The function $\bar{m}(1/M(z))$ is therefore defined and analytic in an annulus-like strip that contains C in its interior. For $z \in C$, $w = 1/M(z)$ lies on $|w| = 1$ or $w\bar{w} = 1$. Hence, on C we have

$$\bar{m}\left(\frac{1}{M(z)}\right) = m\overline{\left(\frac{1}{M(z)}\right)} = \overline{m(M(z))} = \bar{z}.$$

The values of this function on C are therefore $\bar{z}$ and hence,

$$(8.4) \qquad S(z) = \bar{m}\left(\frac{1}{M(z)}\right).$$

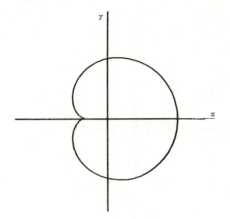

FIG. 8.1 : $z = w + \frac{1}{2}w^2$—Limaçon (cardioid)

Note we have proved once again that the Schwarz Function for C is analytic in an annulus-like strip surrounding C.

The functional equation

$$(8.4') \qquad \bar{M}(S(z)) = 1/M(z)$$

to be solved for a univalent function inside C constitutes the Riemann mapping problem for a region with an analytic boundary. Since volumes have been written on the problem, it is clear that this formulation is deceptively simple.

A similar result can be obtained for the *exterior* mapping function of the curve C.

Example. The limaçon. (See Figs. 8.1, 8.2.) The mapping function is $z = m(w) = w + \sigma w^2$. Hence,

$$M(z) = \frac{-1 + R}{2\sigma}, \qquad R = \sqrt{1 + 4\sigma z}.$$

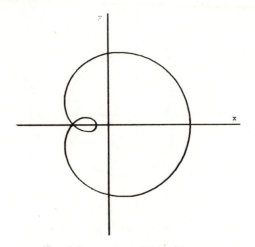

FIG. 8.2 : $z = w + \frac{3}{4}w^2$—Limaçon

Therefore,

$$S(z) = \bar{m}\left(\frac{1}{M(z)}\right) = \frac{2\sigma(R - 1 + 2\mid\sigma\mid^2)}{(R - 1)^2}$$

$$= \frac{(R + 1)(2z + \bar{\sigma}(R + 1))}{4z^2}.$$

Suppose that the analytic arc C in the w plane is mapped onto the arc B in the z-plane by means of the analytic function $w = f(z)$ which is one-to-one conformal in a neighborhood of C. Let $S_B(z)$ and $S_C(w)$ designate the Schwarz Functions of B and C, respectively. Consider the function

(8.5) $$T(w) = \bar{f}(S_B(f^{-1}(w))).$$

Now, for $w \in C$, $f^{-1}(w) = z \in B$; hence, $\overline{T(w)} =$

$\overline{f(S_B(f^{-1}))} = f(\overline{S_B f^{-1}}) = f(f^{-1}(w)) = w$. Thus, $\bar{w} = T(w)$ for $w \in C$, and hence $T(w)$ is the Schwarz Function for C. We can therefore write

$$(8.6) \qquad S_C = \bar{f} S_B f^{-1}; \qquad S_B = \bar{f}^{-1} S_C f,$$

or

$$(8.7) \qquad S_C f = \bar{f} S_B.$$

The functions S_B and S_C will be called (*Hermitian*) *conjugates* under f. Note that this relationship is reflexive, symmetric and transitive. Note also that if S is an involutory function, any conjugate $\bar{f}^{-1} S f$ is also involutory.

As a special instance we have the translation $w = f(z) = z - z_0$, $f^{-1}(w) = w + z_0$, $\bar{f}(w) = z - \overline{z_0}$ which yields

$$(8.8) \qquad S_C(w) = S_B(w + z_0) - \bar{z}_0.$$

As another special case, the rotation $w = f(z) = e^{i\psi}z$, $f^{-1}(w) = e^{-i\psi}w$, $\bar{f}(w) = e^{-i\psi}z$ yields

$$(8.9) \qquad S_C(w) = e^{-i\psi} S_B(e^{-i\psi}w).$$

It is of some interest to see how these algebraic ideas fit in with the conformality of analytic maps. Suppose that C is an analytic arc passing through $z = 0$ with Schwarz Function $S_C(z)$. By (7.8) if ψ_C is the angle the tangent to C at $z = 0$ makes with the x-axis, then $\psi_C = \frac{1}{2} \arg S_C{}'(0)$. Suppose that $f(z)$ is analytic, $f(0) = 0$, $f'(0) \neq 0$, and maps C into the arc B also passing through the origin. If ψ_B is the corresponding angle for B, then $\psi_B = \frac{1}{2} \arg S_B{}'(0)$ where, from (8.6), $S_B = \bar{f}^{-1} S_C f$. Now,

$$S_B{}'(0) = (\bar{f}^{-1})'(S_C(f(0))) \cdot S_C{}'(f(0)) \cdot f'(0)$$

$$= (\bar{f}^{-1})'(0) \cdot S_C{}'(0) \cdot f'(0).$$

Since $f^{-1}f = I$, $(\bar{f}^{-1})'(0) \cdot \bar{f}'(0) = 1$, so that

$$\arg S_B{}'(0) = \arg S_C{}'(0) + \arg \frac{f'(0)}{\bar{f}'(0)}$$

$$= \arg S_C{}'(0) + 2 \arg f'(0).$$

Hence, $\psi_B = \psi_C + \arg f'(0)$, as it should, from elementary conformal mapping. For further material on conformal invariants, see the last section of this chapter.

Analytic continuation. There is a second way of interpreting (8.7). Suppose that $f(z)$ is regular in a region R that contains an analytic arc B as a part of its boundary. Suppose, moreover, that $f(z)$ is continuous along B and, as z approaches points of B, $f(z)$ uniformly approaches points that lie along an analytic arc C.

Consider points z lying exterior to R and near B. For such points S_B is defined and $\overline{S_B(z)}$ yields points interior to R and near B. $f(\overline{S_B(z)})$ is defined yielding points that lie near to C. Finally, $\overline{S_C(f(\overline{S_B(z)}))}$ is defined, again yielding points near to C. Consider now the function $\Phi(z) = \overline{S_C(f(\overline{S_B(z)}))}$. This function is regular analytic exterior to R and near B. For $z \in B$, $\overline{S_B(z)} = z$ and $f(z) \in C$. Hence

$$\Phi(z) = \overline{S_C(f(\overline{S_B(z)}))} = \overline{S_C(f(z))} = f(z).$$

Thus $\Phi(z)$ coincides with f along B. By a familiar argument (see, e.g., Phillips [P6], p. 64), $\Phi(z)$ must be the analytic continuation of f. Thus, the equation

$$(8.7') \qquad f(z) = \bar{S}_C \bar{f} S_B(z)$$

can be regarded as providing *the analytic continuation of f across an analytic arc on which it takes analytic data.* This is a generalization of the classical reflection principle.

Examples. (a) Let B, C be segments of the x-axis;

$S_B = S_C = z$. Then $(8.7')$ becomes $f(z) = \bar{f}(z) = \overline{f(\bar{z})}$. This is the standard *reflection principle*.

(b) Let B and C be the circular arcs $|z - z_0| = r_0$, and $|z - z_1| = r_1$;

$$\overline{S_B(z)} = z_0 + \frac{r_0{}^2}{\bar{z} - \bar{z}_0} = z^*;$$

$$\overline{S_C(z)} = z_1 + \frac{r_1{}^2}{z - \bar{z}_1}.$$

Equation $(8.7')$ becomes $f(z) = z_1 + (r_1{}^2/(\overline{f(z^*)} - \bar{z}_1))$. This formula for analytic continuation plays a considerable role in conformal mapping problems where the region is bounded by circular arcs.

Let R be a simply connected region whose boundary contains a portion of an analytic arc B. Let $w = M(z)$ map R one-to-one conformally onto the unit circle $|w| \leqq 1$. Taking $f(z) = M(z)$ in our previous discussion, $f(z)$ takes values along the unit circle C whose Schwarz Function is $S(z) = \bar{S}(z) = 1/z$. Hence

$$(8.7'') \qquad M(z) = \frac{1}{\bar{M}(S_B(z))}$$

provides the important analytic continuation of M across the analytic arc B.

If the arc C with Schwarz Function S is invariant (curvewise) under a transformation f, then, from (8.7) we have

$$(8.10) \qquad Sf = \bar{f}S.$$

Conversely, suppose one knows that $Sf = \bar{f}S$. The Schwarz Function of the image of C under f is given by $\bar{f}Sf^{-1} = S$, so that C is curvewise invariant.

If, furthermore, f is real on the real axis, then $\bar{f} = f$ so that

$$(8.11) \qquad\qquad Sf = fS$$

and f and S are *permutable functions*.

If an arc with Schwarz Function S is invariant under f and if g maps S into an arc with Schwarz Function T, then this arc is invariant under the map $h = g^{-1}fg$. For by (8.10), $Sf = \bar{f}S$. By (8.6), $T = \bar{g}^{-1}Sg$. Hence $Th = \bar{g}^{-1}Sgg^{-1}fg = \bar{g}^{-1}Sfg = \bar{g}^{-1}\bar{f}Sg = \bar{g}^{-1}\bar{f}\bar{g}\bar{g}^{-1}Sg = \bar{h}T$, and the result follows by (8.10).

Example. As an example of (8.10), consider the group U of Möbius transformations, $f(z) = (az + b)/(cz + d)$, that preserve the unit circle. If $f \sim \left(\begin{smallmatrix} a & b \\ c & d \end{smallmatrix}\right)$ (see 6.22), then $\bar{f} \sim \left(\begin{smallmatrix} \bar{a} & \bar{b} \\ \bar{c} & \bar{d} \end{smallmatrix}\right)$. The Schwarz Function of the unit circle is $S(z) = 1/z \sim \left(\begin{smallmatrix} 0 & 1 \\ 1 & 0 \end{smallmatrix}\right) = E$ or $\left(\begin{smallmatrix} 0 & \lambda \\ \lambda & 0 \end{smallmatrix}\right) = \lambda E$ so that (8.10) yields

$$\lambda \begin{pmatrix} 0 & 1 \\ 1 & 0 \end{pmatrix} \begin{pmatrix} a & b \\ c & d \end{pmatrix} = \begin{pmatrix} \bar{a} & \bar{b} \\ \bar{c} & \bar{d} \end{pmatrix} \begin{pmatrix} 0 & 1 \\ 1 & 0 \end{pmatrix},$$

or

$$\lambda E \begin{pmatrix} a & b \\ c & d \end{pmatrix} = \begin{pmatrix} \bar{a} & \bar{b} \\ \bar{c} & \bar{d} \end{pmatrix} E.$$

This leads to the conditions

$$\lambda \begin{pmatrix} c & d \\ a & b \end{pmatrix} = \begin{pmatrix} \bar{b} & \bar{a} \\ \bar{d} & \bar{c} \end{pmatrix}.$$

From this it follows that we can write

$$f(z) = \frac{az + b}{\bar{b}z + \bar{a}},$$

where we select the normalization $a\bar{a} - b\bar{b} = 1$.

If $\Delta = \left| \begin{smallmatrix} c & d \\ a & b \end{smallmatrix} \right|$, then $\lambda^2 = \overline{\Delta}/\Delta$, so that we may write $\lambda = e^{-i\gamma}$. Select $d = 1$, $c = -\bar{z}_0$; then $a = e^{i\gamma}$, $b = -e^{i\gamma}z_0$ and the transformation appears in the "standard" form

$$w = f(z) = e^{i\gamma} \frac{z - z_0}{1 - z\bar{z}_0}.$$

(See Fig. 8.3.)

A second example.   Let $f(z) = z + a$. The curves C that have complex period a (i.e., if $z \in C$ then $z \pm a \in C$) are invariant under f. Therefore $S(z + a) = S(z) + \bar{a}$ is the functional equation for their Schwarz Function. This is a functional equation of *Abel type*. (The functional equations $gf = g$, $gf = g +$ const. are said to be of *automorphic* and *Abel* type respectively.)

We consider next anti-analytic transformations. For an analytic arc C, we shall designate by $\bar{C}$ the reflection of C in the real axis. Let $S(z)$ be the Schwarz Function of C and suppose that $z \in C$. Then, $\bar{z} = S(z)$. Hence, $\bar{\bar{z}} = z = \overline{S(z)} = \bar{S}(\bar{z})$. Thus $S(z)$ is the Schwarz Function of C if and only if $\bar{S}$ is the Schwarz Function of $\bar{C}$:

$$(8.12) \qquad\qquad S_{\bar{C}}(z) = \bar{S}_C(z).$$

The general anti-analytic transformation $\overline{f(z)}$ may be regarded as f followed by a reflection in the x axis. Combining (8.6) with (8.12), we learn that if the analytic

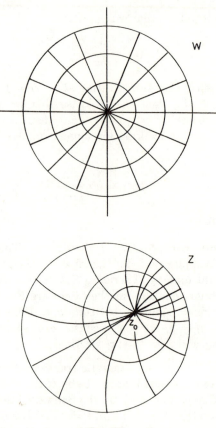

FIG. 8.3

arc C is mapped into the arc B by means of $w = \overline{f(z)}$, then we have

$$(8.12') \qquad S_B = f^{-1} \bar{S}_C \bar{f}.$$

If an arc C with Schwarz Function S is symmetric in the real axis, then $\bar{C} = C$, and hence from (8.12) we obtain

(8.13) $$\bar{S} = S.$$

Since $\bar{S}S = S\bar{S} = I$, we have in this case,

(8.14) $$SS = \bar{S}\bar{S} = S^2 = \bar{S}^2 = I,$$

so that S is *real involutory*.

If the arc C with Schwarz Function S is invariant under an anti-analytic transformation $w = \overline{f(z)}$, then from (8.13), we have $S = f^{-1}\bar{S}\bar{f}$, so that

(8.10') $$\bar{S}\bar{f} = fS.$$

Example. Let the circle $|z - z_0|^2 = r^2$ be invariant under inversion in the unit circle. Here $f(z) = z^{-1}$. Then, by (8.10') this holds if and only if

$$\begin{pmatrix} z_0 & r^2 - |z_0|^2 \\ 1 & -\bar{z}_0 \end{pmatrix} \begin{pmatrix} 0 & 1 \\ 1 & 0 \end{pmatrix} = \lambda \begin{pmatrix} 0 & 1 \\ 1 & 0 \end{pmatrix} \begin{pmatrix} \bar{z}_0 & r^2 - |z_0|^2 \\ 1 & -z_0 \end{pmatrix}.$$

Conclusion: $\lambda = -1$ and $|z_0|^2 = 1 + r^2$ so that the invariant circle is orthogonal to the unit circle.

Suppose we have two analytic arcs with Schwarz Functions S and T. We shall designate the arcs also by S and T. Suppose it is possible to reflect all the points of T in the arc S. The reflected points will lie along an arc U. How is the Schwarz Function of U obtained from S and T? Take a point t on T. Its reflection in S is the point $\overline{S(t)}$. Since this point lies on U, $\overline{\overline{S(t)}} = U(\overline{S(t)})$. Hence, $S(t) = U(\bar{S}(\bar{t}))$. Since t lies along T, $\bar{t} = T(t)$. Hence, $S(t) = U(\bar{S}(T(t)))$, $t \in T$. Therefore $S = U\bar{S}T$. Hence, also,

(8.15) $$T = S\bar{U}S.$$

(8.15') $$U = S\bar{T}S.$$

Certain theorems fall out as very simple algebraic consequences of (8.15′). Suppose that Q is the reflection of R in S, T is the reflection of U in S, and U is the reflection of S in R. Then, T is the reflection of S in Q. For we are given $Q = S\bar{R}S$, $T = S\bar{U}S$, $U = R\bar{S}R$. Then $T = S\bar{R}S\bar{R}S = S\bar{R}S(\bar{S}S)\bar{R}S = Q\bar{S}Q$ and the conclusion follows.

Similarly, if Q is the reflection of R in S, T is the reflection of S in Q, U is the reflection of Q in T, then U is the reflection of R in Q.

Example. The reflection of a circle or a straight line T in a circle S is a circle or a straight line. For the Schwarz Function of S is bilinear while that of T is linear or bilinear. Hence the composite $U = S\bar{T}S$ is linear or bilinear and the conclusion follows. In somewhat more analytic detail, let S be the unit circle and T the circle $|z - z_0| = \rho$. Write

$$S = \begin{pmatrix} 0 & 1 \\ 1 & 0 \end{pmatrix}, \qquad T = \begin{pmatrix} \bar{z}_0 & \rho^2 - |z_0|^2 \\ 1 & -z_0 \end{pmatrix}.$$

Then by (8.15′),

$$U = \lambda S\bar{T}S = \lambda \begin{pmatrix} -\bar{z}_0 & 1 \\ \rho^2 - |z_0|^2 & z_0 \end{pmatrix}.$$

Now select $\lambda = 1/(\rho^2 - |z_0|^2)$, $\bar{w}_0 = -\lambda\bar{z}_0$. Note that $|w_0|^2 + \lambda = \rho^2/(\rho^2 - |z_0|^2)^2$ and select $r = \rho/|\rho^2 - |z_0|^2|$. Then

$$U = \begin{pmatrix} \bar{w}_0 & r^2 - |w_0|^2 \\ 1 & -w_0 \end{pmatrix},$$

and this is the matrix for the Schwarz Function of the

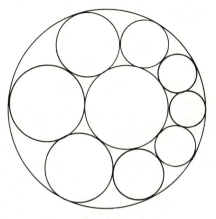

FIG. 8.4

circle $|z - w_0| = r$. If $\rho^2 - |z_0|^2 = 0$, the circle T passes through the origin and the image U is a straight line.

Before leaving the topic of reflections in a circle, and as an application of some of these ideas, we include "Steiner's Porism," one of the most striking theorems in the whole of "visual" plane geometry.

Suppose there are two circles, one inside the other. Suppose further that circles are drawn touching the first two and each other successively as in Figure 8.4. Now, starting from a first circle placed in the ring, there are three possibilities: (1) The ring closes, i.e., the last circle touches the first. (2) The ring closes after a finite number of "go-arounds." (3) The ring never closes. Steiner's Porism asserts that these possibilities are *independent of the starting position of the first circle in the ring*.

To prove this, observe that a bilinear map $z' = (az + b)/(cz + d)$ sends circles into circles (or straight lines). This is so because

$$\frac{az + b}{cz + d} = \frac{bc - ad}{c^2} \frac{1}{(z + d/c)} + \frac{a}{c}.$$

Hence any bilinear map can be built up of the simple transformations $z \to z + \alpha$, $z \to 1/z$, $z \to bz$. Each of these can be seen, e.g., by (8.6), to send circles and straight lines into circles or straight lines.

By continuity, tangent circles go into tangent circles. Now the theorem is perfectly obvious if the two bounding circles are concentric. Hence, it suffices to show that two eccentric circles can be transformed bilinearly into two concentric ones. We shall give a proof of this, exhibiting the fact that this is essentially an eigenvalue problem. We can obviously take the larger circle as the unit circle and place the center of the smaller circle on the real segment $0 < x < 1$.

LEMMA. *Let $0 < r < 1$, $0 < s < 1 - r$. Let*

$$S = \begin{pmatrix} s & r^2 - s^2 \\ 1 & -s \end{pmatrix}, \qquad E = counteridentity = \begin{pmatrix} 0 & 1 \\ 1 & 0 \end{pmatrix}.$$

Then the eigenvalues ρ_1, ρ_2 of SE are real and satisfy $0 < \rho_1 < \rho_2 < 1$. Furthermore

(*) $$(1 - \rho_1)(1 - \rho_2) = s^2.$$

Proof: The characteristic equation of SE is $\rho^2 + \rho(s^2 - r^2 - 1) + r^2 = 0$. Hence $\rho_1 + \rho_2 = 1 + r^2 - s^2$, $\rho_1\rho_2 = r^2$. $(1 - \rho_1)(1 - \rho_2) = s^2$ follows. The discriminant is $\Delta = (1 + r^2 - s^2)^2 - 4r^2$. From the above inequality, $1 - r^2 - s^2 > 2r$. Hence $\Delta > 0$ and ρ_1 and ρ_2 are real and distinct. Now $\rho_2 = \frac{1}{2}(1 + r^2 - s^2 + \sqrt{\Delta}) > r > 0$. Since

$\rho_1\rho_2 = r^2$, $\rho_1 > 0$. Now ρ_1 or $\rho_2 = 1$ is impossible by (*). If $\rho_2 > 1$, then by (*) $\rho_1 > 1$ which contradicts $\rho_1\rho_2 = r^2$.

THEOREM. *Let $0 < s < 1$ and let the circle $S: |z - s| = r$ be contained in interior of the unit circle. Then, the bilinear transformation $A(z)$ with matrix*

$$A \sim \begin{pmatrix} \dfrac{1 - \rho_1}{s} & -1 \\ 1 & \dfrac{-s}{1 - \rho_2} \end{pmatrix}$$

transforms $|z| = 1$ into itself and transforms S into the circle $|z| = \sigma = \sqrt{\rho_1/\rho_2} < 1$.

Proof: In view of the identity (*), we have $EA = -AE$, so by the example after (8.11), $A(z)$ transforms the unit circle into itself. Now, the columns of A are eigenvectors of SE. Hence,

$$SE = A \begin{pmatrix} \rho_1 & 0 \\ 0 & \rho_2 \end{pmatrix} A^{-1}, \qquad SEA = A \begin{pmatrix} \rho_1 & 0 \\ 0 & \rho_2 \end{pmatrix},$$

$$-SAE = A \begin{pmatrix} \rho_1 & 0 \\ 0 & \rho_2 \end{pmatrix}, \qquad SA = -A \begin{pmatrix} \rho_1 & 0 \\ 0 & \rho_2 \end{pmatrix} E,$$

or

$$A^{-1}SA = -\rho_2 \begin{pmatrix} 0 & \sigma^2 \\ 1 & 0 \end{pmatrix}.$$

By (8.6), S is transformed into $|z| = \sigma$.

Let the arc T be curvewise invariant under reflection

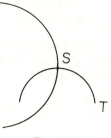

Fig. 8.5

in S. Then, from (8.15′)

$$(8.16) \qquad T = S\bar{T}S$$

or

$$(8.16') \qquad S\bar{T} = T\bar{S}.$$

In other words, the functions S and T are *Hermitian commutative* or Hermitian permutable.

Following the language of reflection in a straight line, we shall say that T is *symmetric* with respect to S.

The converse is also true. If (8.16) holds then T is symmetric with respect to S. For let $z \in T$. Then $\bar{z} = T(z)$. The reflection of z in S is $z^* = \overline{S(z)}$. If $\overline{z^*} = T(z^*)$, it will lie on T. But $\overline{z^*} = \overline{\overline{S(z)}} = S(z)$ and $T(z^*) = T(\overline{S(z)}) = T(\bar{S}(\bar{z})) = $ (by (8.15)) $S(\bar{T}(\bar{z})) = S(\overline{T(z)}) = S(z)$.

The identity (8.16′) can be read as

$$(8.16'') \qquad T\bar{S} = S\bar{T}.$$

This tells us that the *arc T is symmetric with respect to the arc S if and only if the arc S is symmetric with respect to the arc T*. Hence we may speak merely of the symmetry of S and T. (See Fig. 8.5.)

With regard to symmetry, one should observe the fact that an arc S is trivially symmetric with respect to itself. ($S\bar{S} = \bar{S}S = I$.) However, symmetry is *not* an equivalence relationship because transitivity fails.

Special cases of reflections. Let the arc S be symmetric with respect to the line l: $y = (\tan\theta)x$. The Schwarz Function for l is (cf. (3.15)) $T(z) = e^{-2i\theta}z$. Hence from (8.15'),

$$(8.17) \qquad \bar{S}(z) = e^{2i\theta}S(e^{2i\theta}z).$$

If S is symmetric with respect to the real axis, then

$$(8.17') \qquad \bar{S} = S.$$

If S is symmetric with respect to the imaginary axis, then

$$(8.17'') \qquad \bar{S}(z) = -S(-z).$$

If S is symmetric with respect to the unit circle ($T = 1/z$), i.e. is self-inverse, then

$$(8.18) \qquad \bar{S}(z) = \frac{1}{S(1/z)}.$$

If S is symmetric with respect to T while T is symmetric with respect to the real axis ($T = \bar{T}$), then

$$(8.19) \qquad ST = T\bar{S}.$$

If S is symmetric with respect to T while both S and T are symmetric with respect to the real axis, then

$$(8.20) \qquad ST = TS$$

so that S and T are *permutable* functions. (See Fig. 8.6.)

Suppose that T, S, and U are three arcs and the U is the reflection of T in S (or T the reflection of U in S).

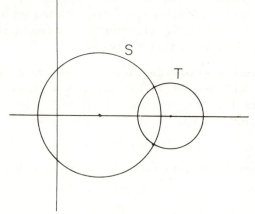

FIG. 8.6

This property is preserved under analytic or anti-analytic maps.

Proof: From (8.15′) we know that $U = S\bar{T}S$. From (8.6) we know that the images under f of T, S, U have the respective Schwarz Functions $\bar{f}^{-1}Tf$, $\bar{f}^{-1}Sf$, $\bar{f}^{-1}Uf$. Now

$$(\bar{f}^{-1}Sf)\overline{(\bar{f}^{-1}Tf)}(\bar{f}^{-1}Sf) = \bar{f}^{-1}S(f\!f^{-1})\bar{T}(\bar{f}\bar{f}^{-1})Sf$$

$$= \bar{f}^{-1}S\bar{T}Sf = \bar{f}^{-1}Uf.$$

Hence the theorem follows from (8.15′). For anti-analytic maps one uses (8.12′).

As a special case we note that:

The symmetry of two arcs is preserved under analytic or anti-analytic maps.

Two circles are symmetric if and only if they are either orthogonal or coincident. For we may take $S(z) =$

$1/z \sim \begin{pmatrix} 0 & 1 \\ 1 & 0 \end{pmatrix} = E,$

$$T(z) = \frac{\rho^2}{z - z_0} + \bar{z}_0 \sim \begin{pmatrix} \bar{z}_0 & \rho^2 - |z_0|^2 \\ 1 & -z_0 \end{pmatrix} = \tau.$$

The condition $S\bar{T} = T\bar{S}$ implies $E\bar{\tau} = \lambda\tau\bar{E}$ for λ constant. With $z_0 \neq 0$, this implies $\lambda = -1$ and $|z_0|^2 = 1 + \rho^2$, the well-known condition for orthogonality. With $z_0 = 0$, we learn that $\rho = 1$ and S and T coincide. One or both of the circles may reduce to straight lines.

More than this, we have the following

THEOREM. *If two analytic arcs are symmetric and have a common point, they are either orthogonal or coincident.*

Proof: Since symmetry is preserved by analytic maps, it is obviously sufficient to prove this when one of the arcs is the real axis. For we merely map one of the arcs onto the x-axis. In the process, the other will be transformed into some analytic arc and orthogonality or coincidence will be preserved. We assume the common point is $z = 0$.

Thus, let there be given an analytic arc with Schwarz Function $S(z)$, symmetric in the real axis. By (8.13) $\bar{S} = S$, so that $\bar{S}'(0) = S'(0)$. Since $|S'| = 1$, $S'(0) = +1$ or -1. If $S'(0) = -1$, S is orthogonal to the x axis. If $S'(0) = 1$, S is tangent to $y = 0$ at $x = 0$. Assume in this case that $S(z) \not\equiv z$. Since the arc is analytic at $z = 0$, we may write it in the form $y = t(x) = a_2x^2 + a_3x^3 + \cdots$, $|x| \leq \sigma$, where not all the a's are 0. Let p be the index of the first nonzero a, so that $y = a_px^p(1 + (a_{p+1}/a_p)x + \cdots)$. For x sufficiently close to 0, we have $1 + (a_{p+1}/a_p)x + \cdots \geq \mu > 0$, so that for p even, the arc lies on one side of the axis, while if p is odd it passes from the 3rd to 1st or

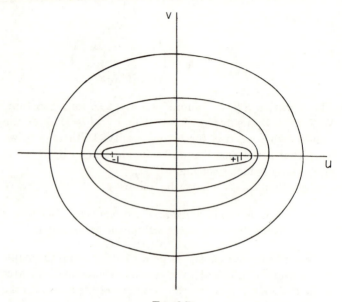

FIG. 8.7

from the 2nd to 4th quadrants. In either case this contradicts the hypothesis that it is symmetric in the x-axis.

Example: Consider the conformal map

$$(8.21) \qquad w = \tfrac{1}{2}(z + z^{-1}).$$

The image of the point $(\rho \cos \theta, \ \rho \sin \theta)$, $\rho > 1$, in the z-plane is the point $(\tfrac{1}{2}(\rho + \rho^{-1}) \cos \theta, \ \tfrac{1}{2}(\rho - \rho^{-1}) \sin \theta)$ in the w-plane. Hence, the circle $|z| = \rho > 1$ maps onto the ellipse $u = \tfrac{1}{2}(\rho + \rho^{-1}) \cos \theta$, $v = \tfrac{1}{2}(\rho - \rho^{-1}) \sin \theta$. Designate this ellipse by $\mathcal{E}_\rho$. As $\rho > 1$ varies, $\mathcal{E}_\rho$ describes a family of confocal ellipses with foci at ± 1 and semiaxes $a = \tfrac{1}{2}(\rho + \rho^{-1})$, $b = \tfrac{1}{2}(\rho - \rho^{-1})$. (See Fig. 8.7.) The image of the unit circle is the interval $-1 \leq u \leq 1$ traced from 1 to -1, thence back to 1.

Now let $1 < \rho_1 < \rho_2$ and let $\rho_3 = \rho_2^2/\rho_1$. If C_ρ designates the circle $|z| = \rho$ then C_{ρ_3} is the reflection of C_{ρ_1} in C_{ρ_2}. Hence it follows from the general principle enunciated that $\mathcal{E}_{\rho_3}$ is the reflection of $\mathcal{E}_{\rho_1}$ in $\mathcal{E}_{\rho_2}$.

Iterations of reflections. Suppose we have two analytic arcs with Schwarz Functions S and T. Then (assuming the operations are possible), if we reflect S in T, then by (8.15′) we obtain $T\bar{S}T$; if we reflect T in $T\bar{S}T$, we obtain $T\bar{S}T\bar{T}T\bar{S}T = T\bar{S}T\bar{S}T$, etc. If we reflect T in S we obtain $S\bar{T}S$. If we reflect S in $S\bar{T}S$ we obtain $S\bar{T}S\bar{T}S$, etc. We observe that if the two arcs S and T intersect, then this can be carried out in a neighborhood of the intersection. We therefore obtain the doubly infinite sequence of arcs

$$\cdots S\bar{T}S\bar{T}S, \; S\bar{T}S, \; S, \; T, \; T\bar{S}T, \; T\bar{S}T\bar{S}T, \cdots$$

If these arcs are symbolized by A_n, $n = 0, \pm 1, \pm 2, \cdots$ then A_{n+2} is the reflection of A_n in A_{n+1} and also A_n is the reflection of A_{n+2} in A_{n+1}. If S is the real axis, $S(z) = z$, so that we obtain the sequence

$$\cdots \bar{T}\bar{T}, \; \bar{T}, \; I, \; T, \; TT, \cdots$$

If there is no confusion with multiplicative powers, we can use exponents to designate *functional iteration*. Note that $\bar{T} = T^{-1}$, so that we can write this last sequence of arcs as

$$\cdots T^{-2}, \; T^{-1}, \; I, \; T, \; T^2, \cdots$$

This gives us geometric interpretation of the successive powers (in the sense of functional composition) of a Schwarz Function.

Example. Let $S(z) = \sqrt{\alpha^2 - z^2}$, $T(z) = \sqrt{\beta^2 - z^2}$, $\beta^2 > \alpha^2$. By (5.13′), S and T are arcs of the family $\mathcal{F}$ of rectangular hyperbolas with common asymptotes $y = \pm x$ and eccentricity $e = \sqrt{2}$. They intersect the x-axis at $x =$

α/e, β/e. Since $T\bar{S}T = \sqrt{2\beta^2 - \alpha^2 - z^2}$, the reflection of S in T is also in $\mathfrak{F}$. The nth iterate of this operation yields the hyperbola with Schwarz Function $\sqrt{(n+1)\beta^2 - n\alpha^2 - z^2}$.

THEOREM. *Let S and T be analytic arcs and suppose that U is the reflection of S in T. If a fourth arc V is symmetric in both S and T, it must also be symmetric in U.*

Proof: $U = T\bar{S}T$. For symmetry of V in U we need $V\bar{U} = U\bar{V}$ or $V\bar{T}S\bar{T} = T\bar{S}T\bar{V}$. Now $S\bar{V} = V\bar{S}$ and $T\bar{V} = V\bar{T}$. Therefore $V\bar{T}S\bar{T} = T(\bar{V}S)\bar{T} = T(\bar{S}V)\bar{T} = T\bar{S}V\bar{T} = T\bar{S}T\bar{V}$.

Example. If circle V is orthogonal to both circles S and T then it is orthogonal to the reflection (inversion) of S in T.

The algebra of functional composition; further considerations. The formal algebra that goes with functional composition and functional inversion has certain computational difficulties associated with it, as we have seen at the beginning of this Chapter. In this section we shall discuss methods for easing some of these difficulties. In dealing with the straight line and the circle whose Schwarz Function is bilinear, some of the computation was lightened by Cayley's method of working with 2×2 matrices. The equation $M_{fg} = \lambda M_f M_g$ permits composition to be expressed in terms of matrix multiplication. Can this be extended to more general functions? It can, as we shall now see; the price is a bit stiff.

By a *formal Laurent series* we shall mean a series of the form

$$(8.22) \qquad L = \sum_{-\infty}^{\infty} a_n z^n,$$

where only a finite number of coefficients with negative subscripts are non-zero. We pay no attention to the convergence of (8.22). The sum of two formal Laurent series is found by adding corresponding coefficients while products are defined formally in the usual way. (The coefficients are Cauchy products.) The set of all formal Laurent series constitutes a field. The *formal derivative* of L is defined as

$$(8.23) \qquad L' = \sum_{-\infty}^{\infty} (n + 1) a_{n+1} z^n.$$

One can easily verify that $(L + M)' = L' + M'$, $(L \cdot M)' = L \cdot M' + L' \cdot M$, and $(L^m)' = mL^{m-1} \cdot L'$. Here L^m designates the mth power of L. The *residue* of (8.22) is defined by

$$(8.24) \qquad \operatorname{Res} L = a_{-1}.$$

In the case in which the series for L converges in some circle $|z| < \rho$, this definition coincides with the usual definition $\operatorname{Res} L = 1/2\pi i \int_\Gamma L(z) dz$, $\Gamma: |z| = \rho' < \rho$. A formal Laurent series L is clearly a (formal) derivative if and only if $\operatorname{Res} L = 0$. For each power m, *positive, zero* and *negative*, define coefficients $a_n^{(m)}$ by means of

$$(8.25) \qquad L^m = \sum_{n=-\infty}^{\infty} a_n^{(m)} z^n.$$

By a *formal power series* we shall mean an expression of the form

$$(8.26) \qquad R = \sum_{n=0}^{\infty} a_n z^n.$$

No attention is paid to the convergence of (8.26). The formal power series form an *integral domain*. We shall designate it by $\mathcal{R}$. Consider also the set of formal power series of the form

$$(8.27) \qquad P = \sum_{n=1}^{\infty} b_n z^n \quad \text{with} \quad b_1 \neq 0.$$

Designate this set by $\mathcal{P}$. Composition is defined by formal substitution and formal rearrangement. Thus,

$$(8.28) \qquad RP = \sum_{n=0}^{\infty} c_n z^n,$$

where

$$(8.29) \qquad c_0 = a_0, \qquad c_n = \sum_{k=1}^{n} a_k b_n^{(k)}.$$

Under composition, $\mathcal{P}$ forms a *noncommutative group* with *unit element*

$$(8.30) \qquad I = 1 \cdot z + 0 \cdot z^2 + 0 \cdot z^3 + \cdots \equiv z.$$

Each element P of $\mathcal{P}$ possesses an inverse P^{-1} satisfying $PP^{-1} = P^{-1}P = I$. To each formal power series $P = \sum_1^{\infty} b_n z^n$ of class $\mathcal{P}$ associate the infinite upper triangular matrix

$$(8.31) \qquad M_P = \begin{pmatrix} b_1^{(1)} & b_2^{(1)} & b_3^{(1)} & \cdots \\ 0 & b_2^{(2)} & b_3^{(2)} & \cdots \\ 0 & 0 & b_3^{(3)} & \cdots \\ & & \cdots\cdots\cdots \end{pmatrix}.$$

Note that M_P is determined completely from its first row through (8.25). The matrix M_I is the infinite unit matrix. Note also that two infinite upper triangular matrices may be multiplied because the sums corresponding to the matrix product contain only a finite number of non-zero terms.

THEOREM. *For any two formal power series P and Q of class $\mathcal{P}$, we have*

$$(8.32) \qquad M_{PQ} = M_P M_Q.$$

Proof: The product of two upper triangular matrices is upper triangular.

By (8.29), the first row of $M_P M_Q$ yields the coefficients of the first power of PQ. Since $(PQ)^2 = (PQ) \cdot (PQ) = P^2Q$, it follows from (8.29) that the second row of $M_P M_Q$ yields the coefficients of $(PQ)^2$. In $P^3Q = (PQ)^3$, the third row yields the coefficients of $(PQ)^3$, etc.

A particular instance of (8.32) comes by selecting $Q = P^{-1}$, in which case

$$(8.33) \quad M_P M_{P^{-1}} = M_{P^{-1}} M_P = M_{PP^{-1}} = M_{P^{-1}P} = M_I = I.$$

Hence we obtain the important identity

$$(8.34) \qquad M_{P^{-1}} = (M_P)^{-1}.$$

Before obtaining more identities, we pause to make an application to the Schwarz Function. Let the arc C pass through the origin $z = 0$. Then, in a neighborhood of the origin we may expand $S(z)$ in the series $S(z) = b_1 z + b_2 z^2 + \cdots$ which is of class $\mathcal{P}$. Each Schwarz Function (of class $\mathcal{P}$) can therefore be represented by the matrix M_S. Since $\overline{P^2} = \bar{P}^2$, etc., $M_{\bar{S}} = \bar{M}_S$. Therefore *all the results on functional composition relating to Schwarz Functions may be expressed in this way as theorems about the matrices M_S*. For example, $\bar{S}S = S\bar{S} = I$ becomes

$$(8.35) \qquad M_S \bar{M}_S = I.$$

Example. Consider the circle $C: |z - 1| = 1$. Its Schwarz Function is $S(z) = z/(z - 1) = -(z + z^2 + \cdots)$ and its associated matrix is

$$G = \begin{pmatrix} -1 & -1 & -1 & -1 & \cdots \\ 0 & 1 & 2 & 3 & \cdots \\ 0 & 0 & -1 & -3 & \cdots \\ 0 & 0 & 0 & 1 & \cdots \\ & & & \cdots\cdots\cdots \end{pmatrix}$$

The identity (8.35) specializes to $G^2 = I$. All the binomial identities embodied in this relationship for the Pascal triangle are therefore aspects of Schwarzian reflection in a circle. As another example, note that the identity (8.1), where $S = \bar{f}f^{-1}$, becomes

$$(8.36) \qquad M_S = \bar{M}_f (M_f)^{-1}.$$

Example. Consider the cubic curve (see Fig. 8.8)

$$\begin{cases} x = t(t - 2) \\ y = t(t - 1)(t - 2) \end{cases}$$

at $t = 0$. We have $z = f(t) = (-2 + 2i)t + (1 - 3i)t^2 + it^3$. Using the matrix identity (8.36), the expansion of the Schwarz Function at $t = 0$ was computed as

$$S(z) = iz - (.25 + .25i)z^2 + (.125 + .1875i)z^3$$

$$- (.0625 + .171875i)z^4 + (.015625 + .166015625i)z^5$$

$$+ (.02734375 + .1606445312i)z^6$$

$$+ (-.0703125 + .1512451172i)z^7 + \cdots$$

Remarks on computation. In a computer language (such as APL) that has excellent matrix implementation, Equations (8.31, 32, 34) constitute a very convenient way for programming functional composition and inversion. The execution is not necessarily the most economical, however. It is often required to deal with power series whose coefficients are complex and therefore complex arithmetic must be provided for. The well-known association $a + bi \leftrightarrow \left(\begin{smallmatrix} a & b \\ -b & a \end{smallmatrix}\right)$ preserves sums, products and scalar products, and it has been found convenient in place of (8.31) to use the matrix

$$\tilde{M}_P = \begin{bmatrix} \begin{pmatrix} \alpha_1{}^1 & \beta_1{}^1 \\ -\beta_1{}^1 & \alpha_1{}^1 \end{pmatrix} & \begin{pmatrix} \alpha_2{}^1 & \beta_2{}^1 \\ -\beta_2{}^1 & \alpha_2{}^1 \end{pmatrix} & \cdots \\ \begin{pmatrix} 0 & 0 \\ 0 & 0 \end{pmatrix} & \begin{pmatrix} \alpha_2{}^2 & \beta_2{}^2 \\ -\beta_2{}^2 & \alpha_2{}^2 \end{pmatrix} & \cdots \\ \begin{pmatrix} 0 & 0 \\ 0 & 0 \end{pmatrix} & \begin{pmatrix} 0 & 0 \\ 0 & 0 \end{pmatrix} & \cdots \\ & \cdots\cdots & \end{bmatrix},$$

where $\alpha_j{}^k = \mathrm{Re}\, b_j{}^{(k)}$, $\beta_j{}^k = \mathrm{Im}\, b_j{}^{(k)}$. The matrix $\tilde{M}_P$ is upper triangular, blockwise.

THEOREM. *Let* $P = \sum_{n=1}^{\infty} b_n z^n$ *and* $P^{-1} = \sum_{n=1}^{\infty} c_n z^n$; *then*

$$(8.37) \qquad c_n{}^{(k)} = \frac{k}{n} b_{-k}^{(-n)}, \qquad n, k = 1, 2, \cdots.$$

Proof: In view of (8.34) we need only to show that

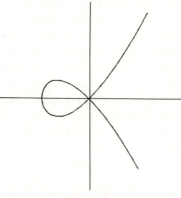

FIG. 8.8

the upper triangular matrix $C = (c_{mn})$ with elements $c_{mn} = (m/n)b_{-m}^{(-n)}$ $n \geq m$ is a right (and hence the) inverse for the matrix M_P. Write $D = (d_{mn}) = M_P C$. D is upper triangular for it is the product of two such. Now for $m = 1, 2, \cdots, d_{mm} = b_m{}^{(m)} \cdot b_{-m}^{(-m)} = b_1{}^m \cdot b_1{}^{-m} = 1$. If $n > m$, then

$$d_{mn} = \sum_{k=m}^{n} b_k{}^{(m)} c_{kn} = \frac{1}{n} \sum_{k=m}^{n} k b_k{}^{(m)} b_{-k}^{(-m)}$$

$$= \text{Res} \left((P^{(m)})' P^{(-n)} \right) = m \, \text{Res} \left(P^{(n-1)} P' P^{(-n)} \right)$$

$$= \frac{m}{m-n} \text{Res} \left(P^{(m-n)} \right)' = 0$$

since it is the residue of a derivative.

The Schwarz Function $S(z) = b_1 z + b_2 z^2 + \cdots$ satisfies $S^{-1} = \bar{S}$; hence

$$(8.38) \qquad \frac{k}{n} b_{-k}^{(-n)} = \bar{b}_n{}^{(k)}, \qquad n, k = 1, 2, \cdots.$$

In the special case $k = 1$, (8.38) becomes

$$(8.39) \qquad n \bar{b}_n b_{-1}^{(-n)} = 1$$

or

$$(8.40) \qquad \bar{b}_n = \frac{1}{n} \text{Res } (S^{(-n)})$$

which embraces the identities (8.3b).

THEOREM (Lagrange-Bürmann). *Let* $R = \sum_{n=0}^{\infty} a_n z^n$ *and let* P *belong to class* $\mathcal{P}$ *and have the inverse series* P^{-1}. *Then*

$$(8.41) \qquad RP^{-1} = a_0 + \sum_{n=1}^{\infty} \frac{1}{n} \text{Res } (R' \cdot P^{(-n)}) z^n.$$

Proof: $M_{RP^{-1}} = M_R M_{P^{-1}} = M_R[(k/n) b_{-k}^{(-n)}]$. Now the coefficients of the series RP^{-1} apart from the 0th are the elements of the first row of $M_{RP^{-1}}$. Writing $RP^{-1} = a_0 + \sum_{n=1}^{\infty} d_n z^n$, we have

$$d_n = \sum_{k=1}^{n} a_k \frac{k}{n} b_{-k}^{(-n)} = \frac{1}{n} \text{Res } (R' \cdot P^{(-k)}), \qquad n = 1, 2, \cdots.$$

If the series for R and P are both regular analytic in some disc $|z| \leqq \rho$, and hence have more than formal meaning, then

$$\text{Res } R' \cdot P^{(-k)} = \frac{1}{2\pi i} \int_{\Gamma} \frac{R'(z) \, dz}{[P(z)]^k},$$

where Γ is a closed curve contained in $|z| \leqq \rho$ and surrounding $z = 0$. Hence,

$$(8.42) \quad RP^{-1}(z) = R(0) + \sum_{n=1}^{\infty} z^n \frac{1}{2\pi in} \int_{\Gamma} \frac{R'(z)\,dz}{[P(z)]^n}$$

$$= R(0) + \sum_{n=1}^{\infty} z^n \frac{1}{n!} \frac{(n-1)!}{2\pi i}$$

$$\times \int_{\Gamma} \left[\frac{z}{P(z)} \right]^n \frac{R'(z)\,dz}{z^n}$$

or

$$(8.43) \quad RP^{-1}(z) = R(0)$$

$$+ \sum_{n=1}^{\infty} \left\{ \frac{z^n}{n!} \frac{d^{(n-1)}}{dz^{n-1}} \left[R'(z) \left[\frac{z}{P(z)} \right]^n \right]_{z=0} \right\}.$$

As a special instance, take $P = f(t) = at + bt^2 + \cdots$, $R = \bar{f}(t)$. Then by (8.1),

$$(8.44) \quad S(z) = \bar{f}f^{-1} = \sum_{n=1}^{\infty} \frac{z^n}{n!} \left[\frac{d^{(n-1)}}{dt^{(n-1)}} \bar{f}'(t) \left(\frac{t}{f(t)} \right)^n \right]_{t=0}$$

$$= \sum_{n=1}^{\infty} z^n \frac{1}{2\pi in} \int_{\Gamma} \frac{\bar{f}'(t)}{[f(t)]^n}\,dt.$$

Now, for Γ taken inside the region of regularity of f and for sufficiently small z, we may interchange the order of integration and summation and write

$$(8.45) \quad S(z) = \frac{1}{2\pi i} \int_{\Gamma} \bar{f}'(t) \sum_{n=1}^{\infty} \frac{1}{n} \frac{z^n}{[f(t)]^n}\,dt$$

$$= \frac{1}{2\pi i} \int_{\Gamma} \bar{f}'(t) \log \frac{f(t)}{f(t) - z}\,dt.$$

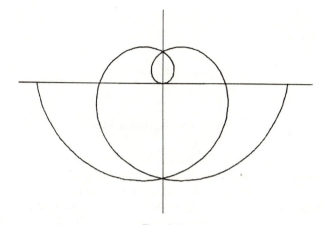

FIG. 8.9

This formula exhibits $S(z)$ as a Cauchy integral in terms of the parameter t on Γ.

Example. The spiral of Archimedes. For fixed $0 < \omega < \infty$, $z = re^{i\theta} = te^{i\omega t} = f(t)$, $0 \leqq t < \infty$. Now $r = t$ and $\theta = \omega t$ so that $r = \theta/\omega$ is the equation of the arc in polar form. We have $\bar{f}'(t)\,[t/f(t)]^n = (1 - i\omega t)\,e^{-i(n+1)\omega t}$. Hence from (8.42),

$$S(z) = 2 \sum_{n=1}^{\infty} \frac{(n+1)^{n-2}(-i\omega)^{n-1}z^n}{(n-1)!} \cdot$$

(See Fig. 8.9).

Conformal invariance of curvilinear angles; the bisection problem and the Schwarz Function. The mapping $z = f(t)$, $0 \leqq t \leqq 1$ was used in (6.4) to define an analytic arc. The inverse mapping f^{-1} transforms the arc into the straight line segment. It follows that any

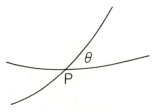

Fig. 8.10—Curvilinear angle

analytic arc may be transformed conformally into any other analytic arc by first referring back to the line segment. More precisely, let two arcs have Schwarz Functions S and T. By (8.1), we have $S = \bar{f}f^{-1}$, $T = \bar{g}g^{-1}$. If we now select $h = fg^{-1}$, we have $T = \bar{g}g^{-1} = \bar{g}\bar{f}^{-1}\bar{f}f^{-1}fg^{-1} = \bar{h}^{-1}Sh$ so that by (8.7) h performs a conformal map of T into S. If a point τ in the interior of T (i.e., $0 < t < 1$) and a point σ in the interior of S are preassigned, then we may obviously find infinitely many conformal transformations h which are regular in a neighborhood of τ, such that $h(\tau) = \sigma$ and h takes T into S. This is known as *Poincaré's local problem of conformal geometry*. (Poincaré showed that the analogous problem for functions of two complex variables lacks a solution in general.) Another way of stating this result is to say that a single analytic arc has no *invariants* with respect to the group of conformal transformations.

The matter is altogether different when one considers the figure formed by two analytic arcs meeting at a point P. Such a figure will be called a *curvilinear angle*. (See Fig. 8.10.) Denote the angle between the arcs by θ. If $\theta = 0$, the figure is often called a *horn angle*. (See Fig. 8.11.) Now if a transformation f is analytic in a neighborhood of P, and $f'(P) \neq 0$, then by the fundamental property of conformality, θ is preserved. Thus, the angle

FIG. 8.11—Horn angle

between two analytic arcs is a conformal invariant. But it turns out that if two angles have the same θ, they are not necessarily equivalent conformally. For example, one can find two analytic arcs intersecting orthogonally which are not conformally equivalent to two perpendicular straight lines.

In terms of Schwarz Functions, the problem is this: Let the first angle have sides with Schwarz Functions P and Q (the curvilinear angle will often be denoted by $\widehat{PQ}$) while the second angle has sides with Schwarz Functions S and T. Desired is an analytic f with

$$(8.46) \qquad \begin{cases} P = \bar{f}Sf^{-1} \\ Q = \bar{f}Tf^{-1}. \end{cases}$$

If S and T are the straight lines $S(z) = \alpha z$, $T(z) = \beta z$, then (8.46) is analogous to a simultaneous diagonalization as the term is used in matrix theory.

Following a lead from matrix theory (that simultaneously diagonalizable matrices must commute), suppose that S and T intersect and are symmetric in each other.

Since the symmetry of two arcs is preserved under analytic maps, if the angle $\widehat{ST}$ is conformally equivalent to the angle $\widehat{PQ}$, it follows that P and Q must be symmetric in each other. To obtain our counterexample, we only need take S and T as two perpendicular lines, P a line, and Q any arc which is orthogonal to, but not symmetric in, P.

Returning to the general situation, let us suppose, as we may, that Q and T are both the real axis with Schwarz Function $Q(z) = T(z) = z$. Take $P(z) = z + \alpha z^2 + \beta z^3 + \cdots$, $S(z) = z + \alpha' z^2 + \beta' z^3 + \cdots$, α, $\alpha' \neq 0$, so that $\widehat{PQ}$ and $\widehat{ST}$ are both horn angles at $z = 0$ with first order contact between P and Q, and S and T. From the second equation in (8.46), $f = \bar{f}$, and therefore from the first equation for conformal equivalence of $\widehat{PQ}$ and $\widehat{ST}$ we must have

$$(8.47) \quad Pf = fS \text{ with } f = a_1 z + a_2 z^2 + a_3 z^3 + \cdots, a_i \text{ real},$$

$$a_1 \neq 0.$$

Inserting the respective power series in (8.47) and expanding up to the third powers, we must have

$$(8.48) \quad Pf = a_1 z + (a_2 + \alpha a_1^2) z^2$$
$$+ (a_3 + 2a_1 a_2 \alpha + \beta a_1^3) z^3 + \cdots$$
$$= a_1 z + (a_1 \alpha' + a_2) z^2$$
$$+ (a_1 \beta' + 2a_2 \alpha' + a_3) z^3 + \cdots$$
$$= fS.$$

From the second coefficient, $\alpha' a_1 = \alpha a_1^2$ so that $a_1 = \alpha'/\alpha$. Inserting this in the third and equating, we get

$$(8.49) \qquad \frac{\beta}{\alpha^2} = \frac{\beta'}{\alpha'^2},$$

and therefore β/α^2 is a *conformal invariant* of the horn angle $\widehat{PQ}$. From (7.23), $\alpha = -ik$, $k = $ curvature, $\beta = -k^2 - (i/3)k'$, $k' = dk/ds$. Therefore, $\beta/\alpha^2 = 1 + (i/3)(k'/k^2)$, so that k'/k^2 is a conformal invariant. Its constancy is a *necessary* condition that two horn angles be conformally equivalent. If the order of contact between P and Q is higher, a similar argument works and it can be shown that every horn angle has one and only one higher conformal invariant.

We turn our attention next to curvilinear angles whose magnitude $\theta = (p/q)\pi$, p, q integers. Let the Schwarz Functions of the sides be T and S. Reflect S in T obtaining $T\bar{S}T$ (see 8.15), reflect T in $T\bar{S}T$ obtaining $T\bar{S}T\bar{S}T$, etc. The angle between T and $T\bar{S}T$ is again θ as is that between $T\bar{S}T$ and $T\bar{S}T\bar{S}T$, etc. In this way we obtain a sequence of arcs passing through the common point of S and T. Since all the angles are equal and since θ is a rational multiple of π, after a finite number of steps we arrive at an arc S' whose direction is that of S. The angle $\widehat{S'S}$ is therefore a horn angle. Any conformal invariant of this horn angle must be a conformal invariant of $\widehat{ST}$ and combining this with the above work we find that every curvilinear angle whose magnitude is a rational multiple of π has a conformal invariant.

Intimately related to the problem of the conformal invariance of curvilinear angles is the "bisection" problem. Suppose that the analytic arcs S and U intersect. Find an arc T through their intersection so that U is the reflection of S in T or equivalently S is the reflection of U in T. Such an arc (if it exists) is called a *bisector* of the curvilinear angle $\widehat{SU}$. (See Fig. 8.12.) From (8.15') we have

$$(8.50) \qquad\qquad S = T\bar{U}T.$$

"Multiply" on the right by $\bar{U}$ yielding

$$(8.51) \qquad\qquad S\bar{U} = T\bar{U}T\bar{U}.$$

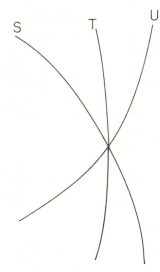

Fig. 8.12

If we now write

$$(8.52) \qquad f = T\bar{U}, \qquad g = S\bar{U},$$

then we get

$$(8.53) \qquad ff = g,$$

to be solved for f. If we were to take U as the x-axis (which we may evidently do), the equation (8.50) becomes

$$(8.54) \qquad TT = S,$$

to be solved for T.

Thus, the bisection problem leads to the determination of a "functional square root." We can clearly pose a similar problem for trisectors and the relevant functional equation is $TTT = S$, etc. Such functional equations are called equations of *Babbage type*.

Assume that the arc S intersects the x-axis U at $z = 0$, and its Schwarz Function has the local expansion

$$(8.55) \qquad S(z) = \sum_{n=1}^{\infty} a_n z^n, \qquad |a_1| = 1.$$

We seek a solution

$$(8.56) \qquad T(z) = \sum_{n=1}^{\infty} b_n z^n, \qquad |b_1| = 1$$

which is also a Schwarz Function. Inserting (8.56) and (8.55) into (8.54) and equating coefficients we obtain successively

$$(8.57) \qquad\qquad\qquad b_1^2 = a_1$$

and for $n = 2, 3, \cdots$

$$(8.58) \quad b_n(b_1^n + b_1) + P_n(b_1, b_2, \cdots, b_{n-1}) = a_n,$$

where P_n is a polynomial which depends on $b_1, \cdots, b_{n-1}$ but not b_n. Now Equation (8.57) has precisely two solutions. Hence, if $b_1^n + b_1 \neq 0$ for $n = 2, 3, \cdots$, then we may solve successively for the coefficients b_n. This will certainly be the case if a_1 is not a root of unity. An expansion (8.56) obtained in this way will be purely formal. This argument shows that if S is any power series and if a_1 is not a root of unity, then (8.54) has *exactly two formal solutions* and nothing is said at this stage whether the formal solutions are, in fact, convergent and whether when S is a Schwarz Function, T will also be a Schwarz Function. Note that from our experience with intersecting lines we anticipate two bisectors, an internal bisector and an external bisector. In a horn angle, if the order of contact of the two curves is exactly 1, it can be shown there is an internal, but no external bisector. If the order of contact is exactly two, there will be two bisectors.

The bisection problem turns out to bè quite difficult and further discussion is postponed to Chapter 15.

Example. Given the two circles $S: |z - s| = s^2$, $U: |z - u| = u^2$, $0 < s < u$ which are tangent at $z = 0$, we can find a circle $|z - t| = t^2$ that bisects the horn angle formed by S and U. Referring to (6.22′) and (8.50), we must have

$$\lambda \begin{pmatrix} s & 0 \\ 1 & -s \end{pmatrix} = \begin{pmatrix} t & 0 \\ 1 & -t \end{pmatrix} \begin{pmatrix} u & 0 \\ 1 & -u \end{pmatrix} \begin{pmatrix} t & 0 \\ 1 & -t \end{pmatrix}$$

$$= \begin{pmatrix} ut^2 & 0 \\ 2ut - t^2 & -ut^2 \end{pmatrix}.$$

Hence $\lambda s = ut^2$, $\lambda = 2ut - t^2$, so that $t = 2us/(u + s)$. In other words, the required bisecting circle has radius t that is the *harmonic mean* of s and u.

As $u \to \infty$, $t \to 2s$, giving us the radius of the bisector of the angle formed by S and the x-axis.

CHAPTER 9

WHAT FIGURE IS THE $\sqrt{-1}$ POWER OF A CIRCLE?

9.1 The $\sqrt{-1}$ power of a circle and spirals. Of course one can make definitions to suit one's own fancy; we shall naturally answer the question in the title of this Chapter in terms of the Schwarz Function.

Suppose that C is an analytic arc (see Fig. 9.1), and let $z \neq 0$ be a variable point on it. Draw the half ray from 0 to z cutting the arc and making an angle $\psi = \psi(z)$ with the arc. The angle ψ will be an analytic function of z. The clinant of the half ray is $\bar{z}/z$ while that of the arc is $S'(z)$. Hence, from (7.9),

$$(9.1) \qquad \frac{S'(z) - (\bar{z}/z)}{S'(z) + (\bar{z}/z)} = -i \tan \psi(z) = \mu(z).$$

Solving for $S'(z)$,

$$(9.2) \qquad S'(z) = \frac{\bar{z}}{z}\left(\frac{1 + \mu(z)}{1 - \mu(z)}\right) = \frac{\bar{z}}{z} e^{-2i\psi(z)}.$$

But along C, $\bar{z} = S(z)$, so that

$$(9.3) \qquad \frac{S'(z)}{S(z)} = \frac{d}{dz} \log S(z) = \frac{e^{-2i\psi(z)}}{z}.$$

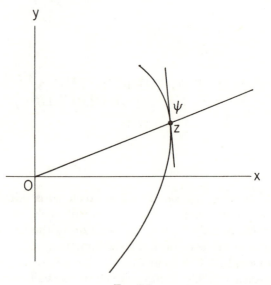

FIG. 9.1

Thus, integrating, the Schwarz Function for C is

$$(9.4) \qquad S(z) = \exp \int e^{-2i\psi(z)} \frac{dz}{z}.$$

Assume that the point $z_0 \in C$. Then $\overline{z_0} = S(z_0)$, and we get

$$(9.5) \qquad S(z) = \overline{z_0} \exp \int_{z_0}^{z} e^{-2i\psi(z)} \frac{dz}{z}.$$

Suppose next that $\psi(z) = \psi = \text{constant}$, $0 \leqq \psi < \pi$. Let

$$(9.6) \qquad \omega = e^{-2i\psi}.$$

Then $S(z)$ in (9.5) takes the form

$$(9.7) \qquad S(z) = \overline{z_0} \exp \left(\omega \log \left(z/z_0\right)\right) = (\overline{z_0}/z_0^{\omega}) z^{\omega}.$$

If, in particular, we set $\psi = \pi/4$, $z_0 = 1$, we obtain the equiangular spiral of Bernoulli of constant angle $\psi = 45°$ passing through $z = 1$. (See Fig. 9.2.) In this case, $\omega = -i$ so that

$$(9.8) \qquad S(z) = 1/z^i$$

is the Schwarz Function of this spiral. The simple form is striking. For $\psi = \pi/2$, $z_0 = 1$, $\omega = -1$, we obtain the unit circle with $S(z) = 1/z$, as we already know. We may therefore say that the 45° spiral of Bernoulli passing through $z = 1$ is "the ith power of the unit circle."

If ω is any complex number on the unit circle, then the "ωth power of the unit circle" is an equiangular spiral. For $\omega = 1$, the spiral degenerates to the unit circle while for $\omega = -1$, it is the real axis. Since $(z^{-i})^i = z$ and $(z^i)^i =$

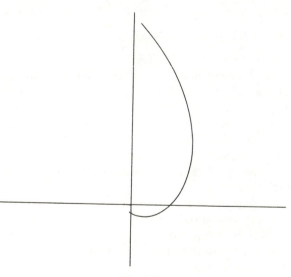

Fig. 9.2

z^{-1}, we have the sequence of figures: unit circle $\rightarrow$ 45° spiral $\rightarrow$ real axis $\rightarrow$ 135° spiral $\rightarrow$ unit circle, under the ith powering of the Schwarz Function. The cycle can also be interpreted geometrically in terms of successive reflections according to the scheme

$$\cdots I, S, SS, \cdots$$

as explained in the previous section (see p. 71). Similar cycles arise when $\omega = e^{-2\pi i r}$, r = rational.

Formula (9.5) gives us the Schwarz Function of an arc in terms of the angle ψ. But the Schwarz Function for the Bernoulli spirals may also be computed expeditiously by means of conformal transformation and (8.6) in the following manner. In the z-plane the line $l: y = \lambda x$ has the Schwarz Function $S(z) = Az$ with $A = (1 - i\lambda)/(1 + i\lambda)$. The image of l under the map $w = f(z) = e^z$ is the spiral $\rho = e^{\theta/\lambda}$. Since $\bar{f}(z) = e^z$ and $f^{-1}(z) = \log z$ we have

$$(9.9) \quad S(z) = \exp{(A \log z)} = z^A, \qquad |A| = 1.$$

Let $|\omega_0| = |\omega_1| = 1$. Then $S_0(z) = z^{\omega_0}$ and $S_1(z) = z^{\omega_1}$ are the Schwarz Functions of two Bernoulli spirals that pass through $z = 1$. Since $\bar{S}_0(z) = z^{\bar{\omega}_0}$, $\bar{S}_1(z) = z^{\bar{\omega}_1}$, we have $S_0(\bar{S}_1(z)) = (z^{\bar{\omega}_1})^{\omega_0} = z^{\omega_0 \bar{\omega}_1}$, and $S_1(\bar{S}_0(z)) = (z^{\bar{\omega}_0})^{\omega_1} = z^{\bar{\omega}_0 \omega_1}$. Now if $\bar{\omega}_0 \omega_1 = \omega_0 \bar{\omega}_1$, then by (8.15) the two spirals are symmetric in each other. This will occur if $\arg \omega_1 - \arg \omega_0 = \pi$. For example, the spirals $\bar{z} = z^i$ and $\bar{z} = z^{-i}$ are symmetric (i.e., Schwarzian reflections of each other).

The spiral $\bar{z} = z^\omega$, $|\omega| = 1$, is curvewise invariant under any conformal map $f(z) = z^\alpha$, α real. For in this case, $\bar{f}(z) = z^\alpha$ and $(z^\alpha)^\omega = (z^\omega)^\alpha$ implies $S\bar{f} = fS$. This implies curvewise invariance by (8.10).

We remark that the spirals $\bar{z} = z^\omega$, $|\omega| = 1$ are the *only* "powers of the unit circle" that represent arcs. For if

$S(z) = (1/z)^p$ is the Schwarz Function of an arc, $\bar{S}S = I$, hence $(z^{-p})^{-\bar{p}} = z^{p\bar{p}} \equiv z$ so that $\mid p \mid = 1$.

The spiral of Bernoulli is also of considerable interest in iteration theory for it provides a prototype behavior of the iteration of an analytic function in a neighborhood of a fixed point. See Chapter 15.

9.2 **Curves invariant under Möbius transformations.** These notions can be brought to bear on the study of curves that are invariant under Möbius transformations. As a preliminary, let us seek curves with Schwarz Function $S(z) = Az^\omega$ that are invariant under $f(z) = \mu z$, $\mu \neq 0$. By (8.10), we must have $A(\mu z)^\omega = \bar{\mu}Az^\omega$, so that independently of A, $\mu^\omega = \bar{\mu}$. This implies that

$$(9.9a) \qquad \omega = \frac{\log \bar{\mu}}{\log \mu} = \frac{\overline{\log \mu}}{\log \mu}.$$

Distinguish three cases:

(1) μ real. Hyperbolic case: dilations. From (9.9a), $\omega = 1$, and the straight lines through the origin, $S(z) = Az$, $\mid A \mid = 1$, are invariant.

(2) $\mid \mu \mid = 1$. Elliptic case: rotations. Here $\omega = -1$ and the circles $S(z) = A/z$, $A > 1$ are invariant.

(3) $\mid \mu \mid \neq 1$, μ not real. Loxodromic case. Then $\mid \omega \mid = 1$, $\omega \neq 1$, $\omega \neq -1$, and the spirals $S(z) = (\bar{z}_0/z_0^\omega)z^\omega$ are invariant.

Now let the general Möbius transformation $w = f(z) = (az + b)/(cz + d)$ be represented by the matrix $M = \begin{pmatrix} a & b \\ c & d \end{pmatrix}$, $\det M \neq 0$. Suppose that the eigenvalues of M are λ_1 and λ_2 and are distinct. Then we may diagonalize M by a G: $G^{-1}MG = \Lambda = \begin{pmatrix} \lambda_1 & 0 \\ 0 & \lambda_2 \end{pmatrix}$ or $M = G\Lambda G^{-1}$. The transformation represented by Λ is $\Lambda(z) = \mu z$ with

$\mu = \lambda_1/\lambda_2$. The invariant curves under M are the images under G of the invariant curves of Λ. In terms of Schwarz Functions, if S is invariant under Λ, the curve with Schwarz Function

$$(9.9b) \qquad\qquad T = \bar{G}SG^{-1}$$

is invariant under M. For

$$TM = \bar{G}SG^{-1}G\Lambda G^{-1} = \bar{G}S\Lambda G^{-1} = \bar{G}\bar{\Lambda}SG^{-1}$$
$$= \bar{G}\bar{\Lambda}\bar{G}^{-1}\bar{G}SG^{-1} = \bar{M}T.$$

CASE 1: μ real. $S \sim \left(\begin{smallmatrix} A & 0 \\ 0 & 1 \end{smallmatrix}\right)$, $|A| = 1$. The straight lines through the origin are invariant under Λ. The members of this pencil intersect at $z = 0$, $z = \infty$. The image of this pencil under G is an elliptic pencil of circles, passing through z_1 and z_2, the images of 0 and ∞ under G. Since, with an obvious notation, $M(z_1, z_2) = G\Lambda G^{-1}(z_1, z_2) = G\Lambda(0, \infty) = G(0, \infty) = (z_1, z_2)$, these are the two fixed points of M.

CASE 2: $|\mu| = 1$, $S \sim \left(\begin{smallmatrix} 0 & A \\ 1 & 0 \end{smallmatrix}\right)$, $A > 0$. The concentric circles around the origin are now invariant under Λ. The points $z = 0$, $z = \infty$ are the "point circles" of this pencil. The image of this pencil under G constitutes a hyperbolic pencil of curves whose point circles are the fixed points of M.

CASE 3: $|\mu| \neq 1$, not real. (Here we cannot use matrices for functional composition with S.) $S(z) = Az^{\omega}$, $A = \bar{z}_0/z_0^{\omega}$, $|\omega| = 1$, $\omega \neq 1, -1$. Write

$$G = \begin{pmatrix} \gamma_1 & \gamma_2 \\ \\ \gamma_3 & \gamma_4 \end{pmatrix}, \qquad G^{-1} = \begin{pmatrix} \Gamma_1 & \Gamma_2 \\ \\ \Gamma_3 & \Gamma_4 \end{pmatrix}.$$

The invariant curves are *loxodromes*, and their Schwarz

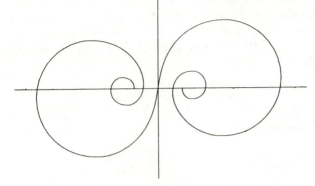

Fig. 9.3

Function is given by

$$(9.9\text{c}) \quad T(z) = \frac{\bar{\gamma}_1 u + \bar{\gamma}_2}{\bar{\gamma}_3 u + \bar{\gamma}_4}, \qquad u = A \left(\frac{\Gamma_1 z + \Gamma_2}{\Gamma_3 z + \Gamma_4} \right)^\omega,$$

(see Fig. 9.3).

9.3 **Differential equations.** To see the spirals against a wider backdrop, we go over to the theory of autonomous differential equations. A 2×2 system of real first order autonomous differential equations can be written in the form

$$(9.10) \quad \begin{cases} \dot{x} = f(x, y) \\ \dot{y} = g(x, y). \end{cases}$$

In conjugate coordinates this becomes

$$(9.11) \quad \begin{cases} \dot{z} = f(x, y) + ig(x, y) = \phi(z, \bar{z}) \\ \dot{\bar{z}} = f(x, y) - ig(x, y) = \overline{\phi(z, \bar{z})} = \bar{\phi}(\bar{z}, z). \end{cases}$$

We suppose here that f and g and hence ϕ are analytic functions of the indicated variables. The second equation in (9.11) is of course redundant. Suppose that C is an arc $z = h(t)$, $t_1 \leqq t \leqq t_2$ which is a solution of (9.11). Along C we have

$$(9.12) \quad S'(z) = \frac{d\bar{z}}{dz} = \frac{\dot{\bar{z}}}{\dot{z}} = \frac{\bar{\phi}(\bar{z}, z)}{\phi(z, \bar{z})} = \frac{\bar{\phi}(S(z), z)}{\phi(z, S(z))}.$$

Therefore

$$(9.12') \qquad\qquad S'(z) = \frac{\bar{\phi}(S(z), z)}{\phi(z, S(z))}$$

is a first order differential equation for the Schwarz Functions of the solutions of (9.10) or (9.11).

If the function ϕ is *separable* in z and $\bar{z}$, i.e., can be written in the form $\phi(z, \bar{z}) = \psi(z)\chi(\bar{z})$ with ψ and χ analytic, then the integral curves of (9.10) have a simple interpretation. In this case,

$$(9.13) \qquad\qquad \frac{dS}{dz} = \frac{\bar{\psi}(S)\bar{\chi}(z)}{\psi(z)\chi(S)},$$

so that

$$(9.14) \quad \int_{S_0}^{S} \frac{\chi(S)}{\bar{\psi}(S)}\, dS = \int_{z_0}^{z} \frac{\bar{\chi}(z)}{\psi(z)}\, dz, \qquad \bar{z}_0 = S(z_0) = S_0.$$

Introduce the function

$$(9.15) \qquad\qquad H(z) = \int_{z_0}^{z} \frac{\bar{\chi}(z)}{\psi(z)}\, dz.$$

Then from (9.14), $\bar{H}(S) - \bar{H}(S_0) = H(z) - H(z_0)$, so that assuming H^{-1} can be found,

$$(9.16) \quad S(z) = \bar{H}^{-1}(H(z) - H(z_0) + \bar{H}(S_0))$$
$$= \bar{H}^{-1}(H(z) - H(z_0) + \overline{H(z_0)}).$$

Write $u_0 = H(z_0)$, $\bar{u}_0 = \overline{H(z_0)}$. From (8.8), the function $l(z) = z - u_0 + \bar{u}_0$ is the Schwarz Function of the line $y = \operatorname{Im} u_0$. From (9.16), $S = \bar{H}^{-1}lH$, so that by (8.6), the solutions of (9.10) or (9.11) are all images under H^{-1} of the family of parallel lines $y = $ constant.

Example. $\dot{z} = z^2\bar{z}$. Here $\psi(z) = z^2$, $\chi(z) = z$, $H(z) = \log z$. $H^{-1}(z) = e^z$. The image under e^z of $y = c$ is $y' = (\tan c)x'$. The differential equation in real form is

$$\begin{cases} \dot{x} = x(x^2 + y^2) \\ \dot{y} = y(x^2 + y^2) \end{cases} \quad \text{so that} \quad \frac{dy}{dx} = \frac{y}{x}$$

and the same conclusion follows.

The 2×2 real first order *linear* autonomous system is

$$(9.17) \quad \begin{cases} \dot{x} = \alpha x + \beta y \\ \dot{y} = \gamma x + \delta y \end{cases} \quad \text{or} \quad \begin{pmatrix} \dot{x} \\ \dot{y} \end{pmatrix} = Q\begin{pmatrix} x \\ y \end{pmatrix}, \quad Q = \begin{pmatrix} \alpha & \beta \\ \gamma & \delta \end{pmatrix}.$$

Now $\begin{pmatrix} z \\ \bar{z} \end{pmatrix} = M\begin{pmatrix} x \\ y \end{pmatrix}$, $M = \begin{pmatrix} 1 & i \\ 1 & -i \end{pmatrix}$ so that in conjugate coordinates, the system becomes

$$(9.18) \quad \begin{pmatrix} \dot{z} \\ \dot{\bar{z}} \end{pmatrix} = P\begin{pmatrix} z \\ \bar{z} \end{pmatrix} \quad \text{with} \quad P = MQM^{-1}.$$

The matrix P can be written in the form

$$(9.19) \quad P = \begin{pmatrix} A & B \\ \bar{B} & \bar{A} \end{pmatrix} \quad \begin{aligned} A &= \tfrac{1}{2}((\alpha + \delta) + i(\gamma - \beta)) \\ B &= \tfrac{1}{2}((\alpha - \delta) + i(\gamma + \beta)). \end{aligned}$$

We note that P and Q are similar, hence their eigenvalues are the same. The invariants are

$$\Delta = (\alpha - \delta)^2 + 4\beta\gamma = p^2 - 4q = (A - \bar{A})^2 + 4B\bar{B}$$

$$= 4(\mid B \mid^2 - (\operatorname{Im} A)^2),$$

$$q = \det Q = \alpha\delta - \beta\gamma = A\bar{A} - B\bar{B} = \mid A \mid^2 - \mid B \mid^2,$$

(9.20)

$$p = -\operatorname{trace} Q = -(\alpha + \delta) = -(A + \bar{A}) = -2 \operatorname{Re} A,$$

$$\lambda_1 = \tfrac{1}{2}(-p + \Delta^{1/2}), \qquad \lambda_2 = \tfrac{1}{2}(-p - \Delta^{1/2}).$$

The point $z = 0$ is a singular point of the system (9.10) or (9.11) and the classification of the singularity can be made completely in terms of the eigenvalues λ_1 and λ_2. It goes as follows:

  (i) Roots real, unequal, same sign ($\Delta > 0$, $q > 0$). Nodal point.
 (ii) Roots real, unequal, opposite sign ($\Delta > 0$, $q < 0$). Saddle point.
(iii) Roots double, Rank of $Q - \lambda I$ is 0 ($\Delta = 0$, $q \neq 0$, $A = $ real, $B = 0$). Star point.
(iv) Roots double. Rank of $Q - \lambda I$ is 1. Nodal point.
 (v) Roots purely imaginary ($\Delta < 0$, $p = 0$). Vortex.
(vi) Roots complex ($\Delta < 0$, $p \neq 0$). Spiral point.

(See Figs. 9.4–9.9.)

From (9.12′) the differential equation for the Schwarz Functions of the solutions of (9.18) is

$$(9.21) \qquad S'(z) = \frac{\bar{B}z + \bar{A}S}{Az + BS}$$

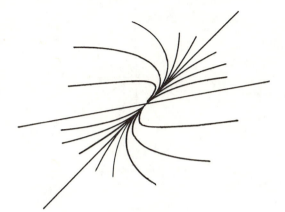

FIG. 9.4—Case (i)

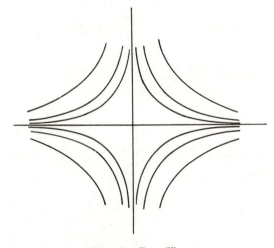

FIG. 9.5—Case (ii)

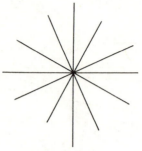

FIG. 9.6—Case (iii)

FIG. 9.7—Case (iv)

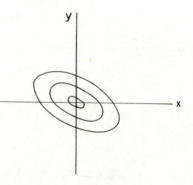

FIG. 9.8—Case (v)

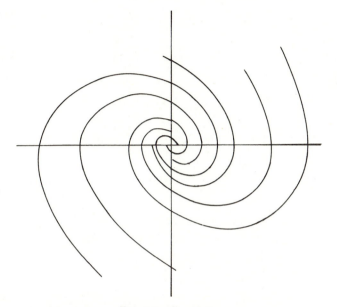

FIG. 9.9—Case (vi)

or equivalently

$$(9.22) \qquad \begin{pmatrix} \dot{z} \\ \dot{S} \end{pmatrix} = \begin{pmatrix} A & B \\ \bar{B} & \bar{A} \end{pmatrix} \begin{pmatrix} z \\ S \end{pmatrix} = P \begin{pmatrix} z \\ S \end{pmatrix}.$$

The solution of (9.22) can be written down in the usual way in terms of the eigenvalues and eigenvectors of P.

Example. $A = 1 + i$, $B = 0$, $\Delta = -4$, $p = -2 \neq 0$. This is case (vi), the spiral point. From (9.21) we have

$$(9.23) \qquad \frac{dS}{dz} = \frac{1 - i}{1 + i} \frac{S}{z} = -i \frac{S}{z}.$$

Hence $S(z) = \text{const } z^{-i}$, and we can recover (9.8).

CHAPTER 10

PROPERTIES IN THE LARGE OF THE SCHWARZ FUNCTION

Given an analytic arc, the global properties of its Schwarz Function $S(z)$ are of considerable interest. The first observation to make, derived from an inspection of the Schwarz Functions we have displayed explicitly in Chapter 5, is that with the exception of the straight line, all the Schwarz Functions have singularities. This is true, as we shall now prove. We base our proof upon the identity (6.11) backed up by a theorem of Pólya on the composition of entire functions.

THEOREM. *Suppose that $f(z)$, $g(z)$, $h(z)$ are entire functions related by*

$$(10.1) \qquad f(z) = g(h(z)).$$

Suppose further that $h(0) = 0$. Let $F(r)$, $G(r)$, $H(r)$ denote respectively the maximum moduli of f, g, h, in the circle $|z| \leqq r$. Then there exists a number c: $0 < c < 1$ independent of g, h, and r such that

$$(10.2) \qquad F(r) \geqq G(cH(\tfrac{1}{2}r)).$$

THEOREM. *The Schwarz Function of an analytic arc C is an entire function of z if and only if C is a straight line.*

Proof: If C is a straight line then $S(z) = Az + B$ and

hence is entire. Suppose, now, that C is not a straight line and $S(z)$ is entire. We may assume that C passes through the origin and hence $S(0) = 0$. (If not perform the translation $w = z - z_0$ which according to (8.8) changes the Schwarz Function to $S_1(w) = S(w + z_0) - \bar{z_0}$. Now $S_1(w)$ is entire if and only if $S(z)$ is.) Thus, we may write $S(z) = a_1 z + a_2 z^2 + \cdots, |z| \leqq \rho$, where at least one coefficient $a_k \neq 0$ with $k \geqq 2$. If $S(z)$ is entire, $\bar{S}(z)$ is entire and since $\bar{S}(z) = \overline{S(\bar{z})}$, both functions have the same maximum modulus function, $M(r)$.

Since $\bar{S}(S(z)) \equiv z$, by Pólya's theorem there exists a constant c, $0 < c < 1$ such that $M(cM(r/2)) \leqq r$. By the Cauchy estimate, $|a_k| r^k \leqq M(r)$ for all r. Hence, $M(r/2) \geqq |a_k| r^k/2^k$, so that once again,

$$r \geqq M\left(cM\left(\frac{r}{2}\right)\right) \geqq |a_k| c^k \left(M\left(\frac{r}{2}\right)\right)^k$$

$$\geqq |a_k| c^k |a_k|^k \left(\frac{r}{2}\right)^{k^2} = \text{const. } r^{k^2}.$$

Since $k \geqq 2$, we obtain a contradiction by allowing $r \to \infty$.

When can the Schwarz Function be rational? The answer is provided by the next theorem.

THEOREM. *The Schwarz Function of an analytic arc C is a rational function of z if and only if C is an arc of a circle or a straight line.*

Proof. The circle $|z - z_0|^2 = \rho^2$ possesses the Schwarz Function $S(z) = (\rho^2/(z - z_0)) + \bar{z_0}$. Conversely, suppose that $S(z) = R(z) = P(z)/Q(z)$ where $P(z) = a_0 z^m + \cdots$, $Q(z) = b_0 z^n + \cdots$, $a_0 b_0 \neq 0$, $(P, Q) = 1$. Since we have identically $\bar{R}(R(z)) = z$,

$$(10.3) \qquad \frac{\bar{a}_0 R^m + \bar{a}_1 R^{m-1} + \cdots}{\bar{b}_0 R^n + \bar{b}_1 R^{n-1} + \cdots} = z,$$

or,

$$(10.4) \quad (\bar{a}_0 R^m + \bar{a}_1 R^{m-1} + \cdots) = z(\bar{b}_0 R^n + \bar{b}_1 R^{n-1} + \cdots).$$

CASE 1. Suppose $m > n$. Then,

$$\bar{a}_0 P^m + \bar{a}_1 Q P^{m-1} + \cdots = Q^m(\bar{b}_0 R^n + \bar{b}_1 R^{n-1} + \cdots).$$

Thus, $Q \mid \bar{a}_0 P^m$, and since $(P, Q) = 1$, $Q = $ const. Thus, $R = \alpha_0 z^m + \alpha_1 z^{m-1} + \cdots$. Since $\bar{R}(R(z)) = z$, identically, it follows that

$$z = \bar{\alpha}_0(\alpha_0 z^m + \alpha_1 z^{m-1} + \cdots)^m + \bar{\alpha}_1(\alpha_0 z^m + \cdots)^{m-1} + \cdots.$$

Comparing leading coefficients, we find $\alpha_0 \bar{\alpha}_0 z^{m^2} = z$ and so $m = 1$, $\alpha_0 \bar{\alpha}_0 = 1$. Then,

$$(10.5) \quad R(z) = \alpha z + \beta, \quad \alpha \bar{\alpha} = 1, \quad \text{Re } (\bar{\alpha}\beta) = 0.$$

CASE 2. Suppose $m < n$. Then, as before, $Q \mid \bar{b}_0 z P^n$, and since $(P, Q) = 1$, $Q = $ const. or $Q = $ const. z. Since P is of smaller degree than Q, we must actually have $R(z) = c/z$. Now $\bar{R}(R(z)) = z$ implies $c = \bar{c}$. Thus, in this case,

$$(10.6) \qquad R(z) = c/z, \qquad c \text{ real.}$$

CASE 3. Suppose $m = n$. Then

$$\bar{a}_0 P^m + \bar{a}_1 Q P^{m-1} + \cdots = z(\bar{b}_0 P^m + \bar{b}_1 Q P^{m-1} + \cdots).$$

This implies that $\bar{a}_0 - \bar{b}_0 z \equiv 0 \pmod{Q}$. Thus, $Q = cz + d$ and so $P = az + b$ for some a, b, c, d. Writing

$$R(z) \sim \begin{pmatrix} a & b \\ c & d \end{pmatrix},$$

then the identity $\bar{R}(R(z)) = z$ implies

$$\begin{pmatrix} \bar{a} & \bar{b} \\ \bar{c} & \bar{d} \end{pmatrix} \begin{pmatrix} a & b \\ c & d \end{pmatrix} = \lambda I.$$

Thus, we need only to look for matrices

$$M = \begin{pmatrix} a & b \\ c & d \end{pmatrix}$$

such that $\bar{M}M = I$, for we have assumed, as we may, that $\det M = 1$. Since $\bar{M} = M^{-1}$, this implies that

$$\begin{pmatrix} \bar{a} & \bar{b} \\ \bar{c} & \bar{d} \end{pmatrix} = \begin{pmatrix} d & -b \\ -c & a \end{pmatrix}$$

so that $\bar{a} = d$, $\bar{b} = -b$, $\bar{c} = -c$, $\bar{d} = a$. Thus,

$$M = \begin{pmatrix} a & ib \\ ic & \bar{a} \end{pmatrix},$$

where $a\bar{a} + bc = 1$. $R(z)$ is now seen to take the form

$$(10.7) \qquad R(z) = \frac{\rho^2}{z - z_0} + \bar{z}_0,$$

with $\rho = c^{-1}$ and $z_0 = i\bar{a}/c$.

All the solutions of the functional equation $\bar{R}(R(z)) = z$ are given by (10.5), (10.6), and (10.7). Case 1 yields the straight lines in the plane, while Cases 2 and 3 are the proper circles.

We shall now give an alternate proof of these theorems

which brings out a somewhat different aspect of the situation.

LEMMA. *Let $f(z)$ be schlicht (univalent) and entire in $|z| < \infty$. Then $f(z)$ has the form $f(z) = a + bz$.*

Proof: The function $f(z)$ cannot be constant and hence by Liouville's Theorem it is unbounded. Set $f(0) = z^*$. Then the unit disc $|z| \leq 1$ is mapped onto a certain (schlicht) region R containing z^*. Hence for $|z| > 1, f(z)$ lies exterior to R. Consider now the function $g(z) = 1/(f(z) - z^*)$. For $|z| > 1$ it is regular and it is bounded. Since it can be continued up to $z = \infty$, it is regular there also. Since $f(z)$ is unbounded in the exterior of R, it follows that $g(\infty) = 0$, and hence $f(z)$ has a pole at $z = \infty$. Now an entire function with a pole at $z = \infty$ must be a polynomial. Since it is schlicht, $f'(z) \neq 0$ for all z. Therefore, by the Fundamental Theorem of Algebra, $f'(z) = $ constant $= b$ and the conclusion follows.

By a similar argument we have

LEMMA. *Let $f(z)$ be schlicht and meromorphic in $|z| < \infty$. Then $f(z)$ has the form $f(z) = (az + b)/(cz + d)$.*

Suppose now that a Schwarz Function $S(z)$ is entire. Then $\bar{S}$ is entire. Then, by permanence of analytic identities, it follows that $\bar{S}S = I$ must be valid for $|z| < \infty$. Suppose $S(z_1) = S(z_2)$. "Multiply" on the left to obtain $\bar{S}S(z_1) = \bar{S}S(z_2)$ or $z_1 = z_2$. Therefore $S(z)$ is schlicht for $|z| < \infty$ and hence by the first lemma, $S(z)$ has the form $a + bz$.

Similar remarks apply if $S(z)$ is assumed to be meromorphic in $|z| < \infty$.

One consequence of the theorems of this chapter is that

if a Schwarz Function is perturbed slightly, it will not in general remain the Schwarz Function of an arc. For example, $z + \epsilon z^2 = S^*(z)$ with ϵ small is a perturbation of the Schwarz Function $S(z) = z$, but with an $\epsilon \neq 0$, $S^*(z)$ is a polynomial of degree 2 and hence not a Schwarz Function.

In Chapter 13 we consider the question of determining when the Schwarz Function of a closed curve can be meromorphic inside the curve.

CHAPTER 11

DERIVATIVES AND INTEGRALS

Derivatives. A function of x and y, $f(x, y)$ can be written as a function of z, $\bar{z}$ through the identity

$$(11.1) \qquad F(z, \bar{z}) = f\left(\frac{z + \bar{z}}{2}, \frac{z - \bar{z}}{2i}\right).$$

Since

$$\frac{\partial F}{\partial z} = \frac{\partial F}{\partial x}\frac{\partial x}{\partial z} + \frac{\partial F}{\partial y}\frac{\partial y}{\partial z} = \frac{\partial F}{\partial x}\left(\frac{1}{2}\right) + \frac{\partial F}{\partial y}\left(\frac{1}{2i}\right),$$

we have

$$(11.2) \qquad \frac{\partial F}{\partial z} = \frac{1}{2}\left(\frac{\partial F}{\partial x} - i\frac{\partial F}{\partial y}\right).$$

Similarly,

$$(11.2') \qquad \frac{\partial F}{\partial \bar{z}} = \frac{1}{2}\left(\frac{\partial F}{\partial x} + i\frac{\partial F}{\partial y}\right).$$

This suggests that it will be convenient to introduce the operators

$$(11.3) \quad \frac{\partial}{\partial z} = \frac{1}{2}\left(\frac{\partial}{\partial x} - i\frac{\partial}{\partial y}\right), \qquad \frac{\partial}{\partial \bar{z}} = \frac{1}{2}\left(\frac{\partial}{\partial x} + i\frac{\partial}{\partial y}\right).$$

Inversely,

$$(11.4) \quad \frac{\partial}{\partial x} = \frac{\partial}{\partial z} + \frac{\partial}{\partial \bar{z}}, \qquad \frac{\partial}{\partial y} = i\left(\frac{\partial}{\partial z} - \frac{\partial}{\partial \bar{z}}\right).$$

The directional derivative of $f(x, y)$ in the direction α is defined by

$$D_\alpha f(x, y) = \lim_{h \to 0} \frac{f(x + h \cos \alpha, y + h \sin \alpha)}{h}$$

$$= \frac{\partial f}{\partial x} \cos \alpha + \frac{\partial f}{\partial y} \sin \alpha.$$

Using (11.4), one obtains

$$(11.5) \qquad D_\alpha F(z, \bar{z}) = F_z e^{i\alpha} + F_{\bar{z}} e^{-i\alpha}.$$

To obtain directional derivatives along and normal to an analytic arc C with Schwarz Function $S(z)$, we proceed as follows. Let $z_0 \in C$ and let ψ be the angle the tangent to C at z_0 makes with the real axis. By definition,

$$\frac{dF}{ds} = \frac{\partial F}{\partial x} \cos \psi + \frac{\partial F}{\partial y} \sin \psi$$

$$(11.5') \quad \frac{dF}{dn} = \frac{\partial F}{\partial x} \cos (\psi + \pi/2) + \frac{\partial F}{\partial y} \sin (\psi + \pi/2)$$

$$= - \frac{\partial F}{\partial x} \sin \psi + \frac{\partial F}{\partial y} \cos \psi.$$

Employing (11.4) we find

$$(11.6) \quad \frac{dF}{ds} = F_z e^{i\psi} + F_{\bar{z}} e^{-i\psi}; \qquad \frac{dF}{dn} = i(F_z e^{i\psi} - F_{\bar{z}} e^{-i\psi}).$$

In particular, the substitution of $F = z$ and $\bar{z}$ in (11.6) yields

$$(11.7) \quad \frac{dz}{ds} = e^{i\psi} = -i \frac{dz}{dn} ; \qquad \frac{d\bar{z}}{ds} = e^{-i\psi} = i \frac{d\bar{z}}{dn} .$$

Combining this with (11.6) we get

$$(11.8) \quad \frac{dF}{ds} = F_z \frac{dz}{ds} + F_{\bar{z}} \frac{d\bar{z}}{ds} ; \qquad \frac{dF}{dn} = F_z \frac{dz}{dn} + F_{\bar{z}} \frac{d\bar{z}}{dn} .$$

From (7.13),

$$\frac{dz}{ds} = -i \frac{dz}{dn} = \frac{1}{\sqrt{S'(z)}} , \qquad \frac{d\bar{z}}{ds} = i \frac{d\bar{z}}{dn} = \sqrt{S'(z)},$$

so that, finally,

$$\frac{dF}{ds} = F_z \frac{1}{\sqrt{S'(z)}} + F_{\bar{z}} \sqrt{S'(z)}$$

(11.9)

$$\frac{dF}{dn} = i \left(F_z \frac{1}{\sqrt{S'(z)}} - F_{\bar{z}} \sqrt{S'(z)} \right).$$

If $f(x, y)$ is a complex-valued function of x, y we may exhibit its real and imaginary parts: $f(x, y) = u(x, y) + iv(x, y)$. Then,

$$\frac{\partial f}{\partial z} = \tfrac{1}{2}(u_x + iv_x - iu_y + v_y), \quad \frac{\partial f}{\partial \bar{z}} = \tfrac{1}{2}(u_x + iv_x + iu_y - v_y).$$

If now, f is an *analytic* function of z, then by the Cauchy-Riemann equations, $u_x = v_y$, $u_y = -v_x$ so that

$$(11.10) \quad \begin{cases} \dfrac{\partial f}{\partial z} = u_x + iv_x = f'(z) \\[4mm] \dfrac{\partial f}{\partial \bar{z}} = 0. \end{cases}$$

Writing $\overline{f(z)}$ for the anti-analytic* function $u - iv$, we have

$$\frac{\partial \overline{f(z)}}{\partial z} = \frac{\partial \bar{f}(\bar{z})}{\partial z} = 0$$

(11.11)

$$\frac{\partial \overline{f(z)}}{\partial \bar{z}} = \frac{\partial \bar{f}(\bar{z})}{\partial \bar{z}} = \overline{f'(z)} = \bar{f}'(\bar{z}).$$

For an analytic function $f(z)$, (11.9) reduces to

$$(11.12) \qquad \frac{df}{ds} = f'(z) \frac{1}{\sqrt{S'(z)}} = -i \frac{df}{dn} \,.$$

The Jacobian. Let $\phi(x, y)$, $\psi(x, y)$ be real functions of x and y and set $F(z, \bar{z}) = \phi + i\psi$, $\bar{F}(\bar{z}, z) = \phi - i\psi$.

Let

$$(11.13) \quad J = \begin{pmatrix} \phi_x & \phi_y \\ \psi_x & \psi_y \end{pmatrix} = \begin{pmatrix} \phi_z + \phi_{\bar{z}} & i(\phi_z - \phi_{\bar{z}}) \\ \psi_z + \psi_{\bar{z}} & i(\psi_z - \psi_{\bar{z}}) \end{pmatrix}.$$

Then,

$$(11.14) \quad \det J = \frac{\partial F}{\partial x} * \frac{\partial F}{\partial y} = -2i \begin{vmatrix} \phi_z & \phi_{\bar{z}} \\ \psi_z & \psi_{\bar{z}} \end{vmatrix}$$

$$= \begin{vmatrix} \phi_z + i\psi_z & \phi_{\bar{z}} + i\psi_{\bar{z}} \\ \phi_z - i\psi_z & \phi_{\bar{z}} - i\psi_{\bar{z}} \end{vmatrix} = \begin{vmatrix} F_z & F_{\bar{z}} \\ \overline{F_{\bar{z}}} & \overline{F_z} \end{vmatrix}$$

$$= |F_z|^2 - |F_{\bar{z}}|^2,$$

---

* Anti-analytic functions are sometimes called *conjugate analytic* functions. There is an isomorphism between the theories of the two types of function. Thus, there is a Cauchy's theorem for $\int_C f(\bar{z})d\bar{z}$, etc.

and

$$(11.15) \qquad \text{trace } J = \phi_x + \psi_y = F_z + \overline{F_{\bar{z}}}.$$

Example. Let $F(z, \bar{z}) = \bar{z} - S(z)$ where $S(z)$ is the Schwarz Function for an arc C. In this case, $\det J = |S'(z)|^2 - 1$. Along C, then, $\det J = 0$.

If the mapping induced by $w = F(z, \bar{z})$ is *orientation preserving*, then $\det J > 0$. If it is *orientation reversing*, then $\det J < 0$. Note that if w is an analytic function of z, then $F_{\bar{z}} = 0$ so that $\det J = |F_z|^2 > 0$, and the transformation preserves orientation. If w is anti-analytic, then $F_z = 0$ so that $\det J = -|F_{\bar{z}}|^2 < 0$ and the orientation is reversed. This is a good device for remembering which way formula (11.14) goes.

Suppose that $\det J > 0$. Then, $|F_z| > |F_{\bar{z}}|$. We have $|D_\alpha F(z, \bar{z})| = |F_z e^{i\alpha} + F_{\bar{z}} e^{-i\alpha}| = |F_z + F_{\bar{z}} e^{-2i\alpha}|$. Hence it is clear that

$$\max_\alpha |D_\alpha F| = |F_z| + |F_{\bar{z}}|$$

$$\min_\alpha |D_\alpha F| = |F_z| - |F_{\bar{z}}| > 0.$$

One can conclude from this that in the neighborhood of a point at which $\det J > 0$, the mapping $w = F(z, \bar{z})$ transforms infinitesimal circles into infinitesimal ellipses wherein the ratio of the major to the minor axes is $D = (|F_z| + |F_{\bar{z}}|) \div (|F_z| - |F_{\bar{z}}|) \geq 1$. Set $\rho = |F_{\bar{z}}| \div |F_z|$. Then, $D = (1 + \rho)/(1 - \rho)$, $\rho = (D - 1) \div (D + 1)$ with $0 \leq \rho < 1$. When $D = 1$, $d = 0$, the mapping is conformal.

The complex-valued function $\rho(z) = F_{\bar{z}}/F_z$ is known as the *complex dilation* of F, and the related differential equation of *Beltrami type* $F_{\bar{z}} = \rho F_z$ is fundamental to the *theory of quasiconformal mappings,* i.e., mappings for which $\rho < k < 1$ throughout a region.

The relation between the complex dilation and the Schwarz Function is as follows. Let $F(z, \bar{z})$ be defined for z in a region R and suppose that $F(z, \bar{z}) = 0$ along an analytic arc C with Schwarz Function $S(z)$. Then, $F(z, S(z)) = 0$ along C, and if F is analytic in z and $\bar{z}$, is identically 0. Differentiating, $F_z + F_{\bar{z}}S'(z) = 0$, so that

$$S'(z) = - \frac{F_z(z, S(z))}{F_{\bar{z}}(z, S(z))}.$$

Along C, then,

$$S'(z) = - \frac{F_z(z, \bar{z})}{F_{\bar{z}}(z, \bar{z})} = - \frac{1}{\rho(z)}.$$

From (3.6), it follows that the slope λ of C is given by $\lambda = i(F_z + F_{\bar{z}})/(F_z - F_{\bar{z}})$. Since $| S'(z) | = 1$ along C, $| \rho(z) | = \rho = 1$. A non-zero quasi-conformal function can therefore not vanish along an analytic arc.

Higher derivatives. From (11.3), we have for the Laplacian

$$(11.16) \quad \frac{\partial^2 F}{\partial z \partial \bar{z}} = \frac{\partial^2 F}{\partial \bar{z} \partial z} = \frac{1}{4}\left(\frac{\partial^2 F}{\partial x^2} + \frac{\partial^2 F}{\partial y^2}\right) \equiv \frac{1}{4}\,\Delta F.$$

The full elliptic differential operator

$$(11.17) \quad L(F) = F_{xx} + F_{yy} + aF_x + bF_y + cF$$

can be written as

$$(11.17') \quad L(F) = F_{z\bar{z}} + \alpha F_z + \beta F_{\bar{z}} + \gamma F,$$

where

$$\alpha = \tfrac{1}{4}(a + ib), \qquad \beta = \tfrac{1}{4}(a - ib), \qquad \gamma = c/4.$$

The higher Laplace operators are obtained by iterating (11.16):

$$(11.18) \qquad \frac{\partial^{2n} F}{\partial z^n \partial \bar{z}^n} = \frac{1}{4^n} \Delta^n F.$$

The solutions of $\Delta^n F = 0$ are called *harmonic functions* when $n = 1$, *biharmonic functions* when $n = 2$, etc.

The formal integration of $\Delta F = 4\partial^2 F / \partial z \partial \bar{z} = 0$ using (10.10) and (10.11) leads to

$$(11.19) \quad F(z, \bar{z}) = f(z) + g(\bar{z}), \qquad f, g \text{ analytic},$$

as a solution of the Laplace equation. Note that $\text{Re}\,(f(z) + g(\bar{z})) = \frac{1}{2}[f(z) + \bar{g}(z) + \bar{f}(\bar{z}) + g(\bar{z})]$, so that by (11.16) $\text{Re}\,(f(z) + g(\bar{z}))$ is a real harmonic function involving two arbitrary functions. In particular, the real and imaginary parts of an analytic function satisfy Laplace's equation $\Delta F = 0$ and are harmonic.

The integration of $\Delta \Delta F = 16 \partial^4 F / \partial z^2 \partial \bar{z}^2 = 0$ leads to

$$(11.20) \quad F(z, \bar{z}) = \bar{z} f(z) + g(z) + z h(\bar{z}) + k(\bar{z}),$$

f, g, h, k analytic, and is a general solution of the biharmonic equation.

Continuation of harmonic functions satisfying nonlinear but analytic boundary data. Suppose that $u(x, y)$ is a harmonic function which, along an analytic arc C with Schwarz Function $S(z)$, satisfies the relationship

$$(11.21) \qquad \partial u / \partial n = \phi(x, y, u),$$

where ϕ is an analytic function of its arguments. Let v be the harmonic conjugate of u so that $f = u + iv$ is an analytic function. We have $u = \frac{1}{2}(f + \bar{f})$ and by (11.9),

$$\frac{\partial u}{\partial n} = \frac{i}{2}\left(\frac{f'(z)}{\sqrt{S'(z)}} - \overline{f'(z)}\sqrt{S'(z)}\right)$$

$$= \phi\left(\frac{z + \bar{z}}{2}, \frac{z - \bar{z}}{2i}, \frac{f + \bar{f}}{2}\right).$$

Write symbolically, with an analytic Φ,

$$\bar{f}'(\bar{z}) = \overline{f'(z)} = \Phi(z, \bar{z}, f(z), \overline{f(z)}, f'(z)) \text{ along } C.$$

Hence

$$(11.22) \quad \bar{f}'(S(z)) = \Phi(z, S(z), f(z), \bar{f}(S(z)), f'(z)).$$

Set $g(z) = \bar{f}(S(z)) = \overline{f(\overline{S(z)})}$. Suppose $f(z)$ is defined on one side of C. $\overline{S(z)}$ lies on the other side of C so that (11.22) is a first order ordinary differential equation

$$(11.23) \quad g'(z) = \Phi(z, S(z), f(z), g(z), f'(z)).$$

This is to be solved under the initial condition $g(z_0) = \overline{f(z_0)}$, for some $z_0 \in C$. The relationship

$$(11.24) \qquad\qquad f(z) = \bar{g}(S(z))$$

then provides the analytic continuation of f across C.

Application to photoelastic computations. As the object here is to present an application of some of the ideas in this chapter, the treatment of the physics and of the equation of plane elasticity is necessarily perfunctory. When a slab B of transparent elastic material is stressed, it acts as a temporary crystal, and when it is viewed under cross polaroids, a chromatic pattern is seen (Fig. 11.1). At all points at which the principal stress difference $p - q$ is constant, the color is the same, and

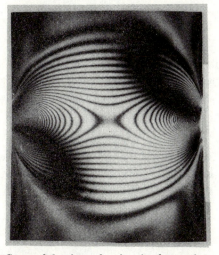

Fig. 11.1—Stressed lamina, showing isochromatics and an isoclinic. (Courtesy Prof. H. Kolsky.)

this gives us an *isochromatic line*. In addition, where the plane of polarization coincides with one of the principal stress directions, a dark spot is obtained; hence, by varying the plane of polarization one can obtain a system of *isoclinic lines* which are the loci of points where the principal stresses make a constant angle α to an initial direction.

Designate the normal stresses in B by σ_{xx}, σ_{yy} and the shear stress by σ_{xy}. Then one has

(A)
$$\begin{cases} \sigma_{xx} - \sigma_{yy} = (p - q) \cos 2\alpha \\ 2\sigma_{xy} = (p - q) \sin 2\alpha. \end{cases}$$

Hence, if one sets

(B)
$$w = \tfrac{1}{2}(\sigma_{yy} - \sigma_{xx} + 2i\sigma_{xy}),$$

one has

(C) $$w = -\tfrac{1}{2}(p - q)e^{-2i\alpha}.$$

The complex quantity w is known as the *conjugate stress deviator* and, as explained above, may be obtained from photoelastic observations. Introduce the *mean normal stress* σ by

(D) $$\sigma = \tfrac{1}{2}(\sigma_{xx} + \sigma_{yy}).$$

By the conditions of compatibility σ is known to be a harmonic function. The equations of equilibrium of an elastic body in a state of plane stress are

(E)
$$\begin{cases} \dfrac{\partial}{\partial x}\,\sigma_{xx} + \dfrac{\partial}{\partial y}\,\sigma_{xy} = 0 \\[3mm] \dfrac{\partial}{\partial x}\,\sigma_{xy} + \dfrac{\partial}{\partial y}\,\sigma_{yy} = 0. \end{cases}$$

Employing the notation of (11.3), and using (B) and (D), these may be written as

(F) $$\frac{\partial \sigma}{\partial z} = \frac{\partial w}{\partial \bar{z}}.$$

Since σ is a real harmonic function in B, it may be written as $\sigma = 2\,\mathrm{Re}\,\Phi(z) = \Phi(z) + \overline{\Phi(z)}$ where Φ is analytic in B. From (11.10), Equation (F) becomes

(G) $$\Phi'(z) = \partial w/\partial \bar{z}.$$

One problem of consequence is this: given w, determine Φ (up to a constant). From (G) one has immediately,

(H) $$\Phi(z) = \Phi(z_0) + \int_{z_0}^{z} \frac{\partial w}{\partial \bar{z}}\, dz.$$

This solution, expressed in real form, is the *shear-difference method*. To use it requires the differentiation of the experimental data embodied in w and is subject to the well-known inaccuracies of numerical differentiation.

Let us integrate (G) with respect to $\bar{z}$. This yields (cf. (11.10))

$$(I) \qquad w = \bar{z}\Phi'(z) + \Psi(z),$$

where Ψ is an arbitrary analytic function in B. If C designates a closed curve in B with Schwarz Function $S(z)$, then one has

$$(J) \qquad w(z) = S(z)\Phi'(z) + \Psi(z), \qquad z \in C.$$

Hence

$$(K) \qquad \int_C w(z)\,dz = \int_C S(z)\Phi'(z)\,dz.$$

Select $C\colon |z - z_0| = r$, $S(z) = (r^2/(z - z_0)) + \bar{z}_0$. Then, by (K) and Cauchy's theorem,

$$(L) \qquad \frac{1}{2\pi i}\int_C w(z)\,dz = \frac{r^2}{2\pi i}\int_C \frac{\Phi'(z)\,dz}{z - z_0} = r^2\Phi'(z_0).$$

Thus, $\Phi'(z)$ is available through the integration of w around circles and $\Phi = \Phi_0 + \int_{z_0}^{z}\Phi'(z)\,dz$. This is known as the *single-circle method*.

Multiply (J) by $(z - z_0)/(z - t)$ where t will be selected interior to C. We obtain

$$(M) \qquad \frac{(z - z_0)}{z - t}\,w(z) = \frac{r^2 + \bar{z}_0(z - z_0)}{z - t}\Phi'(z) + \frac{z - z_0}{z - t}\Psi(z).$$

Integrate this around C obtaining

$$(N) \qquad \frac{1}{2\pi i}\int_C \frac{(z - z_0)w(z)}{z - t}\,dz$$
$$= r^2\Phi'(t) + \bar{z}_0(t - z_0)\Phi'(t) + (t - z_0)\Psi(t).$$

Now take two concentric circles C_1: $|z - z_0| = r_1$, C_2: $|z - z_0| = r_2$, $r_1 < r_2$, and write

(O) $\quad W_j(t) = \dfrac{1}{2\pi i} \displaystyle\int_{C_j} \dfrac{(z - z_0)w(z)}{z - t}\, dz, \qquad j = 1, 2;$

then using (N) two times with $C = C_1$, $C = C_2$ serves to eliminate Ψ:

(P) $\qquad\qquad \Phi'(t) = \dfrac{W_2(t) - W_1(t)}{r_2{}^2 - r_1{}^2}.$

Finally, if one sets

(Q) $\quad \Omega_j(z) = \displaystyle\int_{z_0}^{z} W_j(t)\, dt$

$\qquad\qquad\quad = \dfrac{1}{2\pi i} \displaystyle\int_{C_j} (\zeta - z_0) w(\zeta) \log \dfrac{\zeta - z_0}{\zeta - z}\, d\zeta,$

then

(R) $\qquad\qquad \Phi(z) = \Phi(z_0) + \dfrac{\Omega_2(z) - \Omega_1(z)}{r_2{}^2 - r_1{}^2},$

holding for all z in C_1.

This is known as the *two-circle method* and is more economical than using one circle in that the same values of w on C_1 and C_2 suffice to compute Φ for all z in C_1.

The Cauchy problem for elliptic equations. Suppose that C is an analytic arc and that $f(z)$ and $g(z)$ are two functions that are single-valued and analytic in a region which contains C. The Cauchy problem is to find a solution $F(x, y)$ of $L(F) = 0$ which is twice continuously differentiable in a neighborhood of C and which satisfies

$$F(x, y) = f(z),$$

(11.25) $$(\partial F(x, y)/\partial n) = g(z),$$

$$z = x + iy \in C.$$

Here $L(F)$ is defined in (11.17). The functions f and g are often called the *Cauchy data*. We shall concern ourselves with the analytic continuation of solutions of the Cauchy problem. First, we must build up some background. Let $\mathfrak{R}$ be a region of the $z = x + iy$ plane. We shall assume throughout the remainder of this section that $\mathfrak{R}$ *is simply connected*. By $\overline{\mathfrak{R}}$ we designate the reflection of the region $\mathfrak{R}$ in the x-axis (see 6.12), and by $\mathfrak{R} \times \overline{\mathfrak{R}}$ we designate the region in the space of two complex variables z and z^* such that $z \in \mathfrak{R}, z^* \in \overline{\mathfrak{R}}$.

Suppose that $f(x, y)$ is an analytic function of x and y for $(x, y) \in \mathfrak{R}$. At any point $(x_0, y_0) \in \mathfrak{R}$, $f(x, y)$ has a power series expansion

$$f(x, y) = \sum_{m, n=0}^{\infty} a_{m,n}(x - x_0)^m (y - y_0)^n$$

which is absolutely and uniformly convergent for $|x - x_0| \leqq \rho$, $|y - y_0| \leqq \rho$ and some $\rho > 0$. If we replace x and y by two complex variables $x' + ix''$, $y' + iy''$, with x', x'', y', y'' real, then

$$f(x' + ix'', y' + iy'')$$

$$= \sum_{m, n=0}^{\infty} a_{m,n}(x' + ix'' - x_0)^m (y' + iy'' - y_0)^n$$

converges absolutely and uniformly for $|x' + iy'' - x_0| \leqq \rho$, $|y' + iy'' - y_0| \leqq \rho$ and defines an analytic continuation of $f(x, y)$ into this portion of the space of two complex variables. By a familiar argument, we see that $f(x, y)$ can be continued analytically into a region $\mathbb{Q}$ of the space of two complex variables. The region $\mathbb{Q}$ contains $\mathbb{R}$ and the continuation coincides with $f(x, y)$ when x, y are real.

Make the change of variables

$$(11.26) \qquad z = x + iy$$
$$= (x' + ix'') + i(y' + iy''),$$
$$z^* = x - iy$$
$$= (x' + ix'') - i(y' + iy'').$$

The inverse transformation is

$$(11.27) \qquad \begin{cases} x = \dfrac{1}{2}(z + z^*) \\[2em] y = \dfrac{1}{2i}(z - z^*). \end{cases}$$

(One should not call z and z^* conjugate variables; this will be the case if and only if x and y are both real: $x'' = 0$, $y'' = 0$.) For $f(x, y)$ real analytic, set

$$(11.28) \quad F(z, z^*) = f(x, y) = f(x' + ix'', y' + iy'')$$
$$= f\left(\frac{z + z^*}{2}, \frac{z - z^*}{2i}\right).$$

At (x_0, y_0), we have the local expansion

$$F(z, z^*) = \sum_{m,\, n=0}^{\infty} a_{m,n}(\tfrac{1}{2}(z + z^*) - x_0)^m \left(\frac{1}{2i}(z - z^*) - y_0\right)^n.$$

If $\mathcal{G}$ designates the region $\mathcal{G}: |z - x_0 - iy_0| < \rho/2$, then $\overline{\mathcal{G}}$

will be the region $\overline{\mathcal{G}}$: $|z - x_0 + iy_0| < \rho/2$. Hence, if $(z, z^*) \in \mathcal{G} \times \overline{\mathcal{G}}$, it is clear that $|\frac{1}{2}(z + z^*) - x_0| < \rho$ and $|1/2i(z - z^*) - y_0| < \rho$. Thus, every point $x_0 + iy_0 \in \mathcal{R}$ has a neighborhood $\mathcal{G}$ such that $F(z, z^*)$ is analytic in $\mathcal{G} \times \overline{\mathcal{G}}$.

We shall say that the function $f(x, y)$ belongs to class $V(\mathcal{G})$ if the function of two complex variables $f((z + z^*)/2, (z - z^*)/2i)$ is analytic in $\mathcal{G} \times \overline{\mathcal{G}}$.

Example. Suppose that $u(x, y)$ is a real harmonic function in the simply connected region $\mathcal{G}$. As is well known, one can find a function of a complex variable $f(z)$, analytic in $\mathcal{G}$, such that

$$u(x, y) = \operatorname{Re} f(z) = \tfrac{1}{2}(f(z) + \overline{f(z)}) = \tfrac{1}{2}(f(z) + \bar{f}(\bar{z}))$$
$$= \tfrac{1}{2}(f(x + iy) + \bar{f}(x - iy)).$$

Now $\bar{f}(z^*)$ is analytic for $z^* \in \overline{\mathcal{G}}$. Therefore $U(z, z^*) = u((z + z^*)/2, (z - z^*)/2i) = f(z) + \bar{f}(z^*)$ is analytic for $(z, z^*) \in \mathcal{G} \times \overline{\mathcal{G}}$. Hence $u(x, y)$ is in $V(\mathcal{G})$.

A theorem of Vekua, which we shall now state but not prove, says that this phenomenon of analytic continuability of harmonic functions persists for solutions of the general elliptic equation.

THEOREM. *Let $\mathcal{G}$ be simply connected and suppose that the coefficients of $L(u) = u_{xx} + u_{yy} + au_x + bu_y + cu$ are of class $V(\mathcal{G})$. Then every twice continuously differentiable solution u of $L(u) = 0$ must also be in class $V(\mathcal{G})$.*

Thus, in the strong sense set forth in this theorem, the solution inherits the behavior of the coefficients.

Let us call $\mathcal{G}$ a *fundamental region* of L if it is simply connected and if the coefficients a, b, c are in $V(\mathcal{G})$. For the Cauchy problem, we have the following existence theorem.

THEOREM. *Let $\mathcal{G}$ be a fundamental region for L and suppose that it is conformally symmetric* (see Chapter 8) *with respect to the analytic arc C. Let the Cauchy data $f(z)$, $g(z)$ given on the arc be analytic throughout $\mathcal{G}$. Then, the solution of the Cauchy Problem exists and is twice continuously differentiable in $\mathcal{G}$.*

There is also a converse to this to which we now direct our attention.

THEOREM. *Let $\mathcal{G}$ be a fundamental region for L, and suppose it to be conformally symmetric with respect to the analytic arc C. Let u be a solution of $L(u) = 0$ which is twice continuously differentiable in $\mathcal{G}$. Then both functions*

$$f(z) = u, \qquad g(z) = \partial u/\partial n, \qquad z = x + iy \in C$$

are analytic on C and can be continued analytically into $\mathcal{G}$.

Proof: By Vekua's theorem, $u(x, y)$ has a continuation $U(z, z^*) = u((z + z^*)/2, (z - z^*)/2i)$ which is analytic in $\mathcal{G} \times \overline{\mathcal{G}}$. Now, let $S(z)$ be the Schwarz Function of C. Consider the function $U(z, S(z))$. When $z \in \mathcal{G}$, $S(z) \in \overline{\mathcal{G}}$, so that U is analytic in $\mathcal{G}$. On the other hand, for $z \in C$, $z = \overline{S(z)} = \bar{S}(\bar{z})$ so that $U(z, S(z))$ reduces to $U(z, \bar{z}) = u(x, y)$. Thus, $U(z, S(z))$ is the required continuation of f.

Consider next the function

$$i\left\{ \frac{U_1(z, S(z))}{\sqrt{S'(z)}} - U_2(z, S(z))\sqrt{S'(z)} \right\}$$

$$U_1 = U_z, \qquad U_2 = U_{z^*}.$$

This function is analytic in $\mathcal{G}$, since $S'(z) \neq 0$ in $\mathcal{G}$. On C, $S(z) = \bar{z}$, so that by (11.1) and (11.9) the function in brackets reduces to $\partial u(x, y)/\partial n$, $(x, y) \in C$.

Integrals. Let B be a bounded region in the x, y plane and ∂B its boundary. Then, Green's theorem in two variables tells us that

$$(11.29) \quad \int_{\partial B} Pdy - Qdx = \int\int_B (P_x + Q_y)dxdy.$$

Quite general sufficient conditions on P, Q and B for this to be true will be found, e.g., in Bochner [**B5**], Lehto and Virtanen [**L1**, p. 156].

Consider now the integral

$$(11.30) \quad \frac{1}{2i}\int_{\partial B} \overline{f(z)}g(z)dz = \frac{1}{2i}\int_{\partial B} \bar{f}g(dx + idy),$$

where f and g are analytic functions regular (for simplicity) in $B + \partial B$. Write

$$P = \frac{1}{2}\bar{f}g \qquad Q = \frac{i}{2}\bar{f}g.$$

Then, from (11.4), $P_x = \frac{1}{2}(\partial/\partial z + \partial/\partial \bar{z})\bar{f}g$, $Q_y = -\frac{1}{2}(\partial/\partial z - \partial/\partial \bar{z})\bar{f}g$, so that $P_x + Q_y = \partial/\partial \bar{z}(\bar{f}g) = \partial/\partial \bar{z}(\bar{f}(\bar{z})g(z)) = \overline{f'(z)}g(z)$. From (11.29),

$$(11.31) \quad \int\int_B \overline{f'(z)}g(z)dxdy = \frac{1}{2i}\int_{\partial B} \overline{f(z)}g(z)dz$$

$$= \frac{1}{2i}\int_{\partial B} \bar{f}(\bar{z})g(z)dz.$$

This establishes Green's theorem in a complex analytic form. Suppose now that ∂B consists of a single analytic curve whose Schwarz Function is $S(z)$. Then $\bar{z} = S(z)$ along ∂B so that

$$(11.32) \quad \int\int_B \overline{f'(z)}g(z)dxdy = \frac{1}{2i}\int_{\partial B} \bar{f}(S(z))g(z)dz.$$

In particular, the selection $f(z) \equiv z$ yields

$$(11.33) \qquad \int_B \int g(z) \, dx \, dy = \frac{1}{2i} \int_{\partial B} S(z) g(z) \, dz.$$

THEOREM. *Let f be regular in $B + \partial B$, and let $\bar{f}(S(z)) \ (= \overline{f(\overline{S(z)})})$, be regular in all of B. Then f is identically constant.*

Proof: (Note that for values of z interior to B and close to ∂B, $\overline{S(z)}$ is exterior to B and close to ∂B. Therefore $\bar{f}(S(z))$ is initially known to be regular near C without the further hypothesis.) We have from (11.32) and by Cauchy's theorem,

$$\int_B \int |f'(z)|^2 \, dx \, dy = \frac{1}{2i} \int_{\partial B} \bar{f}(S(z)) f'(z) \, dz = 0.$$

Therefore $f'(z) \equiv 0$ and the conclusion follows.

By selecting $f \equiv z$ we see that $S(z)$ must have a singularity inside B and furthermore that the singularities of S cannot be annihilated by its being composed with a regular function.

The nature of the singularity of $S(z)$ inside B crucially governs the kind of application that can be made. We shall exhibit three types of singularity: (a) An S for a region B with a piecewise analytic boundary; (b) An S which is meromorphic in B; (c) An S which is made single-valued by a branch cut inside B.

(a) We begin with a region in which the boundary and hence the Schwarz Function is only piecewise analytic. Let T be a triangle whose vertices in counterclockwise order are z_1, z_2, z_3. Let $A = \text{area}(T)$. Let the sides of T be designated by T_1, T_2, T_3. Along T_1 we have, from (3.10),

$$\bar{z} = S(z) = S_1(z) = A_1 z + B_1,$$

$$A_1 = \frac{\bar{z}_1 - \bar{z}_2}{z_1 - z_2}, \qquad B_1 = \frac{z_1 \bar{z}_2 - z_2 \bar{z}_1}{z_1 - z_2}.$$

Similarly for the other sides. Now,

$$\int_T \int f''(z)\,dx\,dy = \frac{i}{2} \int_{T_1 + T_2 + T_3} f'(z)\,d\bar{z}$$

$$= \frac{i}{2} \int_{z_1}^{z_2} + \int_{z_2}^{z_3} + \int_{z_3}^{z_1} f'(z)\,d\bar{z}.$$

Now $\int_{z_1}^{z_2} f'(z)\,d\bar{z} = A_1 \int_{z_1}^{z_2} f'(z)\,dz = A_1(f(z_2) - f(z_1))$, and similarly for the other integrals. Therefore,

$$\iint_T f''(z)\,dx\,dy = \frac{i}{2} \{ (A_3 - A_1)f(z_1)$$

$$+ (A_1 - A_2)f(z_2) + (A_2 - A_3)f(z_3) \}$$

Now,

$$A_3 - A_1 = \frac{\bar{z}_3 - \bar{z}_1}{z_3 - z_1} - \frac{\bar{z}_1 - \bar{z}_2}{z_1 - z_2}$$

$$= \frac{1}{(z_1 - z_2)(z_1 - z_3)} \begin{vmatrix} z_1 & \bar{z}_1 & 1 \\ z_2 & \bar{z}_2 & 1 \\ z_3 & \bar{z}_3 & 1 \end{vmatrix}$$

$$= (\text{by } (3.21)) = \frac{-4iA}{(z_1 - z_2)(z_1 - z_3)}.$$

Similar formulas hold for $A_1 - A_2$ and $A_2 - A_3$. Hence,

$$(11.34) \quad \int_T \int f'' dx dy = 2A \left\{ \frac{f(z_1)}{(z_1 - z_2)(z_1 - z_3)} \right.$$

$$\left. + \frac{f(z_2)}{(z_2 - z_1)(z_2 - z_3)} + \frac{f(z_3)}{(z_3 - z_1)(z_3 - z_2)} \right\}.$$

This formula can be put in the alternative forms

$$(11.35) \quad \frac{1}{2A} \int_T \int f''(z) \, dx dy$$

$$= f(z_1, z_2, z_3)$$

$$= \begin{vmatrix} 1 & 1 & 1 \\ z_1 & z_2 & z_3 \\ f(z_1) & f(z_2) & f(z_3) \end{vmatrix} : \begin{vmatrix} 1 & 1 & 1 \\ z_1 & z_2 & z_3 \\ z_1^2 & z_2^2 & z_3^2 \end{vmatrix}$$

$$= \text{2nd divided difference of } f \text{ at } z_1, z_2, z_3.$$

$$(11.36) \quad R(f; z_1) = \frac{(z_1 - z_2)(z_1 - z_3)}{2A} \int_T \int f''(z) \, dx dy,$$

where R is the remainder at z_1 due to linear interpolation to $f(z)$ at z_2 and z_3. (For the relationship between divided differences and interpolation remainders, see, e.g., Davis [**D1**] p. 64.)

Suppose that $f(z_1) = f(z_2) = 0$. Then (11.34) reduces to

$$(11.37) \quad f(z_3) = \frac{(z_3 - z_1)(z_3 - z_2)}{2A} \int_T \int f''(z) \, dx dy$$

$$= \frac{(z_3 - z_1)(z_3 - z_2)}{2}$$

$$\times \text{ Average of } f'' \text{ over } T.$$

Taking absolute values we arrive at the following inequality:

Suppose that $f(z)$ is analytic in a convex region $\mathcal{R}$ containing the points z_1 and z_2. Suppose further that $f(z_1) = f(z_2) = 0$ and $|f''(z)| \leqq 1$ throughout $\mathcal{R}$. Then

$$(11.38) \qquad |f(z)| \leqq \tfrac{1}{2} |z - z_1| |z - z_2|$$

in $\mathcal{R}$.

Example. Let T designate the equilateral triangle with vertices at 1, w, w^2 where $w = \exp(2\pi i/3)$. Then,

$$\int_T \int f'' dx dy = (\sqrt{3}/2)\{ f(1) + wf(w) + w^2 f(w^2) \}.$$

Further identities may be derived. Suppose that P is a polygon with vertices at $z_1, z_2, \cdots, z_n$. Decompose P into triangles and apply (11.34) to each triangle. Then we can find constants $a_1, \cdots, a_n$ depending only upon P such that

$$(11.39) \qquad \int_P \int f'' dx dy = \sum_{j=1}^{n} a_j f(z_j)$$

for all functions f that are regular analytic inside P and continuous in its closure.

Suppose now that k is a fixed integer $\geqq 0$ and that $p(z)$ is any polynomial of degree $\leqq k$. If we set

$$g(z) = \sum_{\alpha=0}^{k} (-1)^{k-\alpha}(k - \alpha + 1)p^{(k-\alpha)}(z)f^{(\alpha)}(z),$$

then it is easily shown by differentiation that

$$g''(z) = p(z)f^{(k+2)}(z).$$

Hence, from (11.39),

$$\int_P \int p(z)f^{(k+2)}(z)\,dx\,dy$$

$$= \sum_{j=1}^{n} a_j g(z_j)$$

$$= \sum_{j=1}^{n} a_j \sum_{\alpha=0}^{k} (-1)^{k-\alpha}(k - \alpha + 1)p^{(k-\alpha)}(z_j)f^{(\alpha)}(z_j).$$

In place of $p(z)$ write $(z - t)^k$, so that

$$\int_P \int (z - t)^k f^{(k+2)}(z)\,dx\,dy$$

$$= \sum_{j=1}^{n} a_j \sum_{\alpha=0}^{k} (-1)^{k-\alpha}(k - \alpha + 1)f^{(\alpha)}(z_j)\,\frac{k!}{\alpha!}\,(z_j - t)^\alpha.$$

This is a "Darboux-type" formula for the double integral.

(b) We consider next a region Q whose Schwarz Function is meromorphic in Q. Take for Q the bicircular quartic (5.16): $r^2 \leqq a^2 + 4\epsilon^2 \cos^2 \theta$,

$$S(z) = \frac{z(a^2 + 2\epsilon^2) + z\sqrt{a^4 + 4a^2\epsilon^2 + 4\epsilon^2 z^2}}{2(z^2 - \epsilon^2)}.$$

The quantity under the radical vanishes when $z = \pm i\sqrt{a^2 + (a^4/4\epsilon^2)}$. Since $a^4/4\epsilon^2 > 0$, and $r = a$ when $\theta = \pi/2, 3\pi/2$, these points lie outside the curve Q. Hence, inside the curve, $S(z)$ can be defined as a single-valued analytic function with simple poles $z = \pm\epsilon$. Now, $\iint_Q f(z)\,dx\,dy = (\pi/2\pi i)\int_{\partial Q} S(z)f(z)\,dz$. Since the only singularities of $S(z)$ are simple poles at $z = \pm\epsilon$, we need only evaluate the residues at $\pm\epsilon$. For $z = \epsilon$ we have

$$(z - \epsilon)S(z) = \frac{\epsilon(a^2 + 2\epsilon^2) + \epsilon\sqrt{a^4 + 4a^2\epsilon^2 + 4\epsilon^4}}{2(2\epsilon)}$$

$$= \frac{\epsilon a^2 + 2\epsilon^3 + \epsilon a^2 + 2\epsilon^2}{4\epsilon} = \frac{a^2}{2} + \epsilon^2.$$

A similar result holds for $z = -\epsilon$. Hence,

$$(11.40) \quad \int_Q \int f(z)\,dxdy = \pi\left(\frac{a^2}{2} + \epsilon^2\right)(f(\epsilon) + f(-\epsilon)).$$

The selection $f(z) \equiv z^{2m}$, and evaluation of the left-hand integral in polar coordinates yields the integral identity

$$(11.41) \quad \int_0^{2\pi} e^{i2m\theta}(a^2 + 4\epsilon^2\cos^2\theta)^{m+1}d\theta$$

$$= (2m + 2)\pi(a^2 + 2\epsilon^2)\epsilon^{2m}.$$

For $m = 0$ we obtain

$$(11.42) \quad \int_Q \int dxdy = \text{area}(Q) = \pi(a^2 + 2\epsilon^2).$$

More generally, we have for $n \geqq 0$ by (11.32),

$$(11.43) \quad \int_Q \int \bar{z}^n f(z)\,dxdy = \frac{1}{2i(n + 1)}\int_{\partial Q} S^{n+1}(z)f(z)\,dz.$$

For simplicity, write

$$(11.44) \quad N(z) = z/2((a^2 + 2\epsilon^2) + \sqrt{a^4 + 4a^2\epsilon^2 + 4\epsilon^2 z^2})$$

so that $S^{n+1}(z) = [N^{n+1}(z)]/[(z - \epsilon)^{n+1}(z + \epsilon)^{n+1}]$ and this function is regular inside ∂Q, except at $z = \pm\epsilon$ where it has poles of order $n + 1$. Hence,

$$(11.45) \quad \int_Q \int \bar{z}^n f(z)\, dx dy$$

$$= \frac{\pi}{(n+1)!} \frac{n!}{2\pi i} \int_{\partial Q} \frac{N^{n+1}(z)}{(z - \epsilon)^{n+1}(z + \epsilon)^{n+1}} f(z)\, dz$$

$$= \frac{\pi}{(n+1)!} \left\{ \frac{d^n}{dz^n} \left(\frac{N^{n+1}(z)f(z)}{(z + \epsilon)^{n+1}} \right) \right|_{z=\epsilon}$$

$$+ \frac{d^n}{dz^n} \left(\frac{N^{n+1}(z)f(z)}{(z - \epsilon)^{n+1}} \right) \bigg|_{z=-\epsilon} \right\}$$

$$= \frac{\pi}{(n+1)!} \left[a_{n0} f(\epsilon) + a_{n1} f'(\epsilon) + \cdots + a_{nn} f^{(n)}(\epsilon) \right.$$

$$\left. + b_{n0} f(-\epsilon) + b_{n1} f'(-\epsilon) + \cdots + b_{nn} f^{(n)}(-\epsilon) \right],$$

where the constants a_{nk} and b_{nk} are independent of f and may be obtained explicitly by expanding the above bracket.

(c) We consider finally a case where the Schwarz Function is regular in a slit region. We take the ellipse $\mathcal{E}$:

$$\frac{x^2}{a^2} + \frac{y^2}{b^2} \leq 1, \qquad a > b$$

$$S(z) = \frac{a^2 + b^2}{a^2 - b^2} z + \frac{2ab}{b^2 - a^2} \sqrt{z^2 + b^2 - a^2}.$$

The first term of S is a regular function in $\mathcal{E}$. Hence

$$(11.46) \quad \int_{\mathcal{E}} \int f(x)\,dx\,dy = \frac{ab}{i(b^2 - a^2)} \int_{\partial\mathcal{E}} \sqrt{z^2 + b^2 - a^2}\, f(z)\,dz$$

$$= \frac{ab}{b^2 - a^2} \int_{\partial\mathcal{E}} \sqrt{a^2 - b^2 - z^2}\, f(z)\,dz.$$

The function $\sqrt{a^2 - b^2 - z^2}$ is single-valued in the plane slit along $-\sqrt{a^2 - b^2} \leqq x \leqq \sqrt{a^2 - b^2}$. Hence, we may shrink the curve $\partial\mathcal{E}$ until it coincides with the slit traversed twice; the lower edge from $-\sqrt{a^2 - b^2}$ to $\sqrt{a^2 - b^2}$ and the upper edge back. On the first traversing, $dz = dx$ and the radical is $-\sqrt{a^2 - b^2 - x^2}$; on the return, $dz = -dx$ and the radical is $+\sqrt{a^2 - b^2 - x^2}$. Hence,

$$(11.47) \quad \int_{\mathcal{E}} \int f(z)\,dx\,dy$$

$$= \frac{2ab}{a^2 - b^2} \int_{-\sqrt{a^2-b^2}}^{\sqrt{a^2-b^2}} \sqrt{a^2 - b^2 - x^2}\, f(x)\,dx.$$

It is often convenient to place the foci of $\mathcal{E}$ at ± 1. Write $a = \frac{1}{2}(\rho + \rho^{-1})$, $b = \frac{1}{2}(\rho - \rho^{-1})$ and designate by $\mathcal{E}_\rho$ the ellipse with foci at ± 1 and semi-axis sum $a + b = \rho$. Then, $2ab = \frac{1}{2}(\rho^2 - \rho^{-2})$ and

$$(11.48) \quad \int_{\mathcal{E}_\rho} \int f(z)\,dx\,dy = \frac{1}{2}(\rho^2 - \rho^{-2}) \int_{-1}^{+1} \sqrt{1 - x^2}\, f(x)\,dx.$$

The identity (11.48) is related to the orthogonality of the Tschebyscheff polynomials of the second kind, $U_n(z)$, over ellipses $\mathcal{E}_\rho$.

CHAPTER 12

APPLICATION TO ELEMENTARY FLUID MECHANICS

The theory of analytic functions is often applied to plane problems of fluid flow, heat flow, electrostatics, and elasticity. We select fluid flow as a prototype, and show its relationship to the Schwarz Function.

Let D be a region in the complex z-plane. A *velocity field* or a *steady flow* on D comprises two functions

$$(12.1) \qquad u(z) = u(x, y); \qquad v(z) = v(x, y)$$

which provide the x and the y velocity components of a fluid particle located at $z = x + iy$. One way of visualizing the flow is by erecting the vector $\mathbf{v} = u\mathbf{i} + v\mathbf{j}$ at each point z_0. Introduce the *complex velocity* of the flow by means of

$$(12.2) \qquad q(z) = u + iv.$$

The line joining z_0 to $z_0 + q$ has the clinant $\bar{q}/q = (u - iv)/(u + iv)$. Suppose now that

$$(12.3) \qquad p(z) = \phi(z) + i\psi(z)$$

is an analytic function defined in D and is such that $p'(z)$ is single-valued. Suppose further that

$$(12.4) \quad q(z) = u + iv = -\frac{\overline{dp(z)}}{dz} = -\overline{p'(z)} = -\bar{p}'(\bar{z}).$$

135

The flow (u, v) is then called a *potential flow* and $p(z)$ is called its *complex velocity potential*. Every analytic function $p(z)$ with a single-valued derivative gives rise to a potential flow. Conversely, it is well known that any differentiable flow which is divergence- and circulation-free is derivable from a complex potential through (12.4). From (11.10) we have $dp/dz = \phi_x + i\psi_x$ so that

$$(12.5) \quad q = u + iv = -(\overline{dp/dz}) = -\phi_x + i\psi_x.$$

Therefore

$$(12.6) \qquad u = -\phi_x; \qquad v = \psi_x.$$

From the Cauchy-Riemann equations we have, in fact,

$$(12.6') \quad u = -\phi_x = -\psi_y; \qquad v = \psi_x = -\phi_y.$$

The clinant of the velocity vector is

$$\bar{q}/q = (-p'(z))/(\overline{-p'(z)}).$$

Consider now the arc

$$(12.7) \quad S: \quad \psi = \operatorname{Im} p(z) = c, \qquad c = \text{real const.}$$

We have $\operatorname{Im} p(z) = 1/2i(p(z) - \overline{p(z)}) = c$; hence, $p(z) - \bar{p}(\bar{z}) = ci$. If $S(z)$ is the Schwarz Function for the arc S then

$$(12.8) \qquad p(z) - \bar{p}(S(z)) = ci.$$

along S. Hence by a standard uniqueness theorem for analytic functions, (12.8) holds identically in z. Solving for $S(z)$ we have

$$(12.9) \qquad S(z) = \bar{p}^{-1}(p(z) - ci).$$

Differentiating (12.8), we find $p'(z) - \bar{p}'(S(z)) \cdot S'(z) = 0$, so that

$$(12.10) \qquad S'(z) = p'(z)/\bar{p}'(S(z))$$

and hence along S, $S'(z) = p'(z)/\bar{p}'(\bar{z})$. The clinants of S coincide with those of the velocity field so that the curves (12.7) are the *stream lines* of the flow.

The curves

(12.11) $T:$ $\phi = \operatorname{Re} p(z) = c,$ $c = \text{real const.}$,

are called the *velocity equipotentials*. If $T(z)$ designates the Schwarz Functions of these curves, a similar argument shows that

(12.12) $p(z) + \bar{p}(T(z)) = c$

(12.12′) $T(z) = \bar{p}^{-1}(c - p(z))$

(12.12″) $T'(z) = -p'(z)/\overline{p'(z)}$.

Hence by (7.12) *the two families of curves are orthogonal.*

Let ϕ and ψ be the real and imaginary parts of an analytic function $p(z)$. *Then the curves $\phi = $ const. and $\psi = $ constant are symmetric in each other.* For, $S = \bar{p}^{-1}(p - ic)$ and $T = \bar{p}^{-1}(c_1 - p)$. Hence $\bar{T} = p^{-1}(c_1 - \bar{p})$, $\bar{S} = p^{-1}(\bar{p} + ic)$ so that $S\bar{T} = \bar{p}^{-1}(pp^{-1}(c_1 - \bar{p}) - ic) = \bar{p}^{-1}(c_1 - ic - \bar{p})$. Similarly, $T\bar{S} = \bar{p}^{-1}(c_1 - pp^{-1}(\bar{p} + ic)) = \bar{p}^{-1}(c_1 - ic - \bar{p}) = S\bar{T}$.

We shall call the functional equations (12.8) and (12.12) of *conjugate Abel type*. If we set $\Pi(z) = e^{p(z)}$, then the equation (12.8) becomes $\overline{\Pi}S = e^{-ic}\Pi$, which is a functional equation of *conjugate Schroeder type*.

The speed $|q|$ of the fluid particle located at z is given by

(12.13) $|q|^2 = q\bar{q} = p'(z)\overline{p'(z)} = |p'(z)|^2$

and the *stagnation points* of the flow (points of zero velocity) are located at the zeros of $p'(z)$.

Associated with the flow of potential $p(z)$ is the integral

(12.14) $J = -\int_C p'(z)dz$

taken around a closed contour C in D. One has

$$(12.15) \quad J = \int_C (udx + vdy) + i \int_C (-vdx + udy).$$

The two real integrals in (12.15) are called the *circulation* and *the flux* or *divergence* of the flow around the contour C:

$$(12.16) \quad \text{Circulation} = \int_C udx + vdy = -\text{Re} \int_C p'(z)dz,$$

$$(12.17) \quad \text{Divergence} = \int_C -vdx + udy = -\text{Im} \int_C p'(z)dz.$$

Many complex potentials have been studied explicitly in closed form. We exhibit some of the simple ones.

Examples.

(a) *Uniform flow.* $D: |z| < \infty$. $p(z) = \sigma e^{-i\alpha}z$, $\sigma \geqq 0$, α real. $p'(z) = \sigma e^{-i\alpha}$. Clinant of velocity vector: $e^{-2i\alpha}$. Speed is uniformly σ (see Fig. 12.1).

(b) *Doublet or dipole.* $D: |z| > 0$. $p(z) = \sigma e^{i\alpha}/z$, $\sigma > 0$, α real. $\bar{p}^{-1} = (\sigma e^{-i\alpha}/z)$ so that from (12.9) the stream lines have the Schwarz Function

$$S(z) = \frac{ic\sigma e^{-i\alpha}z}{z + ic\sigma e^{i\alpha}}, \qquad c \text{ real constant.}$$

This is a family of circles passing through the origin and whose centers lie on the line $\bar{z} = -e^{-2i\alpha}z$ (see Fig. 12.2).

Complex potentials may be added to one another.

(c) *Motion past a circle, uniform at $z = \infty$.* $D: |z| \geqq a$. $p(z) = \beta z + \gamma/z$ where $\beta = ue^{-i\alpha}$, $\gamma = ua^2 e^{i\alpha}$, u real,

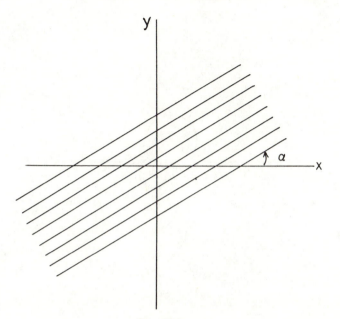

Fig. 12.1

$a > 0$. Since $p'(z) = \beta - \gamma/z^2$, as $z \to \infty$, the flow approaches the uniform flow described in (a). The circle $|z| = a$ has Schwarz Function a^2/z and since $\gamma = \bar{\beta}a^2$, $p(z) - \bar{p}(S(z)) = 0$ so that this circle is a stream line. The stagnation points are determined from $p'(z) = 0$. This yields $z = \pm ae^{i\alpha}$ and these points lie on the circle (see Fig. 12.3).

With appropriate interpretations, $p(z)$ may be selected as multivalued (providing $p'(z)$ is single-valued).

(d) *Source, sink, vortex, source and vortex.* $D: |z| > 0$.

CASE 1. $p(z) = \sigma \log z$, σ real. If $\sigma < 0$ we have a

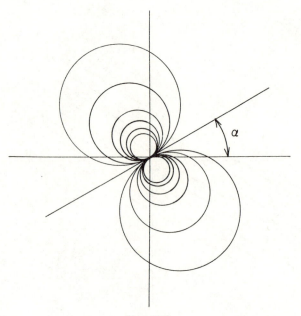

FIG. 12.2

source. If $\sigma > 0$ we have a *sink.* $p'(z) = \sigma/z$ and is single-valued. $q = -\sigma/\bar{z}$. The clinant of the velocity is $\bar{q}/q = \bar{z}/z$ and is that of the radius vector. The velocity is therefore radial. The speed $|q| = |\sigma|/|z|$ and hence approaches infinity as $|z| \to 0$ (see Fig. 12.4).

CASE 2. $p(z) = \sigma i \log z$, σ real. This is a *vortex.* $\bar{p}^{-1}(z) = \exp(iz/\sigma)$ so that the stream lines are given by $S(z) = \bar{p}^{-1}(p - ci) = e^{c/\sigma}/z$, the Schwarz Function for a circle centered at the origin. We have $|q|^2 = |p'(z)|^2 = \sigma^2/|z|^2$, so that the speed is constant along each circle (see Fig. 12.5).

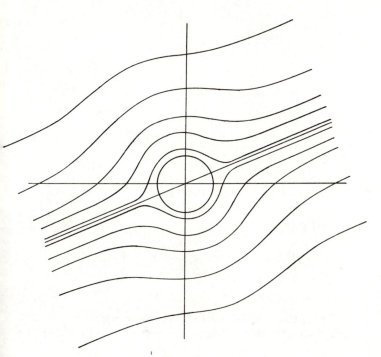

FIG. 12.3

CASE 3. $p(z) = a \log z$, a complex. *Sink or source and vortex*. In this case, $\bar{p}^{-1}(z) = \exp(z/\bar{a})$. Hence $S(z) = \bar{p}^{-1}(p(z) - ci) = \exp((1/\bar{a})(a \log z - ci)) = e^{-ic/\bar{a}}z^{\omega}$ with $\omega = a/\bar{a}$. These are *spirals of Bernoulli* (see Fig. 12.6).

(e) *Vortex line*. $p(z) = \sigma i \log \sin \pi z/a$, $\sigma > 0$, $a > 0$. Vortices of "strength" σ are located at $z = 0, \pm a, \pm 2a, \cdots$. $\bar{p}^{-1}(z) = (a/\sigma) \sin^{-1}(\exp(iz/\sigma))$. The stream lines are

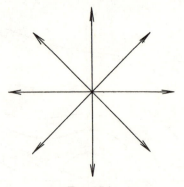

FIG. 12.4

given by

$$S(z) = \bar{p}^{-1}(p - ci) = \frac{a}{\pi}\sin^{-1}\left(e^c \csc \frac{\pi z}{a}\right)$$

(see Fig. 12.7).

The principle of conformal mapping. Suppose that $p(z)$ is a complex potential defined in a region D of the

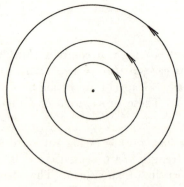

FIG. 12.5

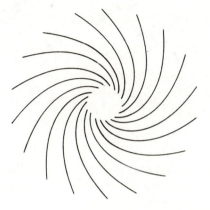

FIG. 12.6

z-plane and has a streamline S with Schwarz Function $S(z)$. Let the analytic function $w = f(z)$ map D one to one conformally onto E in the w-plane, the arc S going into the arc T. Then $pf^{-1}(w)$ is a complex potential in E with streamline T.

To prove this we need only show that $pf^{-1}(w) - \overline{pf^{-1}}(T(w)) = ci$, c real. Now by (8.6), $T = \bar{f}Sf^{-1}$, so we need show that $pf^{-1} - \overline{pf^{-1}}\bar{f}Sf^{-1} = pf^{-1} - \bar{p}Sf^{-1} = ci$. Since S is a streamline for p, $p - \bar{p}S = c'i$, so that applying f^{-1} on the right we obtain the result with $c' = c$.

The circle theorem, extended. If $p(z)$ is any complex potential in a region D and if $S(z)$ is the Schwarz Function of an arc S lying in D, then $\bar{p}(S(z)) = \overline{p(\overline{S(z)})}$ is defined in a neighborhood of S, and

$$h(z) = p(z) + \bar{p}(S(z))$$

is a complex potential for which S is a streamline. For we

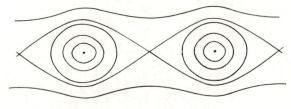

FIG. 12.7

need only show that $h(z) - \bar{h}(S(z)) = ci$. But $\bar{h}(S(z)) = \bar{p}(S(z)) + p(\bar{S}(S(z))) = \bar{p}(S(z)) + p(z)$. Therefore $h(z) - \bar{h}(S(z)) = 0$.

As particular instances, select $S(z) = z$; then $h(z) = p(z) + \bar{p}(z)$, and the x axis becomes a streamline. If we select $S(z) = \frac{1}{z}$, then $h(z) = p(z) + \bar{p}(\frac{1}{z})$ and the unit circle is a streamline. This is known as the *Circle Theorem*.

The formulas of Blasius. Suppose that a flow with complex potential $p(z)$ has a closed streamline C with Schwarz Function $S(z)$. The formulas of Blasius express the force and moment experienced by C as a result of the flow.

Let the local pressure be designated by $\mathfrak{p} = \mathfrak{p}(z)$. If the fluid is incompressible, it is known from Bernoulli's equation that $\mathfrak{p} = a - \frac{1}{2}\rho \mid q \mid^2$, where a is a constant and ρ is the uniform density of the fluid. Select units so that $\rho = 2$. Then $\mathfrak{p} = a - \mid q \mid^2 = a - \mid p'(z) \mid^2$. Designate the differentials of force by dX and dY. We have $dX = -\mathfrak{p}dy$, $dY = \mathfrak{p}dx$,

$$X = \int_C dX, \qquad Y = \int_C dY,$$

so that

$$dX - idY = -i\mathfrak{p}(dx - idy) = -i\mathfrak{p}d\bar{z}.$$

Therefore

$$(12.18)\quad X - iY = -i\int_C \mathfrak{p}(z)d\bar{z} = i\int_C (\,|\,p'(z)\,|^2 - a)\,d\bar{z}$$

$$= i\int_C p'(z)\overline{p'(z)}S'(z)\,dz$$

$$= (\text{by } (12.10))\; i\int_C p'(z)\overline{p'(z)}\frac{p'(z)}{\overline{p'(z)}}\,dz$$

$$= i\int_C (p'(z))^2 dz.$$

Thus,

$$(12.19)\qquad X - iY = i\int_C (p'(z))^2 dz.$$

If M designates the moment of the forces about the origin, then $dM = xdY - ydX = \mathfrak{p}(xdx + ydy) = \mathfrak{p}\,\mathrm{Re}\,(zd\bar{z})$. Therefore

$$(12.20)\qquad M = \int_C dM = \int_C \mathfrak{p}\,\mathrm{Re}\,(zd\bar{z})$$

$$= \mathrm{Re}\int_C (a - |\,p'(z)\,|^2)zd\bar{z}.$$

Now $\int_C zd\bar{z} = -2i\cdot$area of C so that $\mathrm{Re}\int_C zd\bar{z} = 0.$

Hence, as with the forces,

$$(12.21) \qquad M = - \operatorname{Re} \int_C z \mid p'(z) \mid^2 dz.$$

If the potential $p(z)$ has no singularities inside C, the integrals (12.19), (12.21) vanish.

By the *kinetic energy* of the portion of the flow inside C is meant the integral

$$(12.22) \qquad KE = \int_C \int \tfrac{1}{2}\rho \mid q \mid^2 dxdy.$$

Taking $\rho = 2$, we have

$$(12.23) \quad KE = \int_C \int \mid p'(z) \mid^2 dxdy = \frac{1}{2i} \int_C \overline{p(z)} p'(z) dz$$

$$= \frac{1}{2i} \int_C \bar{p}(S(z)) p'(z) dz.$$

To obtain the theorem of Kutta-Joukowski, one assumes that the airfoil C is in a uniform air stream of speed V and circulation k. The complex potential is therefore taken to be

$$p(z) = Ve^{i\alpha} - A \log z + \frac{B}{z} + \cdots, \quad -A = \frac{ik}{2\pi},$$

where $q(z) = B/z + \cdots$ is regular exterior to C and is such that $\lim_{z \to \infty} zq(z) = B \neq \infty$. Then $p'(z) = Ve^{i\alpha} - (A/z) + q'(z)$ so that

$$(p'(z))^2 = V^2 e^{2i\alpha} + \frac{ikVe^{i\alpha}}{\pi z} + \cdots$$

It follows from (12.19) that $X - iY = -2ikVe^{i\alpha}$ and hence

$$(12.24) \qquad X + iY = 2kVe^{i(\pi/2-\alpha)}.$$

From (12.21),

$$(12.25) \qquad M = \text{Re } 4\pi iBVe^{i\alpha}.$$

The relation (12.24) is the theorem of Kutta-Joukowski, while (12.25) is the theorem of Blasius.

CHAPTER 13

THE SCHWARZ FUNCTION AND THE DIRICHLET PROBLEM

By the *Dirichlet Problem* or *First Boundary Value Problem* in two dimensions is meant the following: given a region B in the (x, y) plane and given a function $u(z^*)$ defined for the boundary points z^* of B, find a function $u(z)$ which is harmonic in B (i.e., satisfies $\Delta u = 0$ at interior points of B) and for which $\lim_{z \to z^*} u(z) = u(z^*)$ in some sense. For the purposes of this essay, we limit ourselves to a region B whose boundary is an analytic curve with Schwarz Function $S(z)$ and where the boundary function $u(z^*)$ is continuous.

The solution of the Dirichlet problem is well known when B is the circle: $|z| < R$. It is

$$(13.1) \qquad u(t) = \frac{1}{2\pi} \int_{|z|=R} \frac{R^2 - |t|^2}{|z - t|^2} \, u(z) d\theta.$$

A short calculation shows that if we write $S(t) = R^2/t$ for the Schwarz Function of the circle $|z| = R$, then

149

formula (13.1) can be written in the form

$$(13.2) \quad u(t) = \text{Re} \left\{ \frac{1}{2\pi i} \int_{|z|=R} \frac{u(z)}{z - t} \, dz \right.$$

$$\left. - \frac{1}{2\pi i} \int_{|z|=R} \frac{u(z)}{z - \overline{S(t)}} \, dz \right\}.$$

Since $\overline{S(t)}$ is the image or the reflection of t in the circle $|z| = R$, one speaks of solving the Dirichlet problem by the *Method of Images*. The formula (13.2) can be generalized to certain regions other than the circle. We shall now give an indication of the procedure and the proof. Set

$$(13.3) \quad u(t) = \text{Re} \left\{ \frac{1}{2\pi i} \int_{\partial B} \frac{u(z) \, dz}{z - t} - \frac{1}{2\pi i} \int_{\partial B} \frac{u(z) \, dz}{z - \overline{S(t)}} \right\}$$

$$= \text{Re} \left\{ \frac{1}{2\pi i} \int_{\partial B} \frac{u(z) \, (t - \overline{S(t)})}{(z - t) (z - \overline{S(t)})} \, dz \right\}.$$

In the first place, with $u(z)$ continuous on ∂B, $(1/2\pi i) \int_{\partial B} (u(z) \, dz)/(z - t)$ defines a function $f(t)$ which is a regular analytic function of t throughout the interior of B. Similarly, the integral $(1/2\pi i) \int_{\partial B} (u(z) \, dz)/(z - \overline{S(t)})$ defines a regular analytic function $g(\bar{t})$ of $\bar{t}$ throughout an annulus-like strip in B and bordering on ∂B. It follows from (11.16) that $u(t) = \text{Re} \, (f(t) + g(\bar{t}))$ is a harmonic function. In general, $u(t)$ will not be continuable throughout the interior of B as a single-valued branch of a harmonic function. However, this will be the case if $S(t)$, the Schwarz Function of B, is itself *meromorphic* inside B. The poles of $S(t)$ produce zeros of g.

We show finally that $\lim_{t \to z^*} u(t) = u(z^*)$ as t approaches z^* along a normal to ∂B.

With t inside B, $\overline{S(s)}$ will be outside B. Hence, by Cauchy's theorem,

$$\frac{1}{2\pi i}\int_{\partial B}\frac{dz}{z-t}-\frac{1}{2\pi i}\int_{\partial B}\frac{dz}{z-\overline{S(t)}}=1,$$

so that

$$\mathrm{Re}\left\{\frac{1}{2\pi i}\int_{\partial B}\frac{u(z^*)\,dz}{z-t}-\frac{1}{2\pi i}\int_{\partial B}\frac{u(z^*)\,dz}{z-\overline{S(t)}}\right\}=u(z^*),$$

and hence,

$$(13.4)\quad u(t)-u(z^*)$$

$$=\mathrm{Re}\left\{\frac{1}{2\pi i}\int_{\partial B}(u(z)-u(z^*))\frac{t-\overline{S(t)}}{(z-t)(z-\overline{S(t)})}\,dz\right\}.$$

Taking absolute values,

$$(13.5)\quad |u(t)-u(z^*)|$$

$$\leqq\frac{1}{2\pi}\int_0^L|u(z)-u(z^*)|\frac{|t-\overline{S(t)}|}{|z-t|\,|z-\overline{S(t)}|}|dz|,$$

where $L=$ length of ∂B. Let us parametrize the points of ∂B by the arc length $0\leqq s\leqq L$, with $z\leftrightarrow s$ and $z^*\leftrightarrow s^*$. Since $u(s)$ is continuous, we may set $\max_{0\leqq s\leqq L}|u(s)-u(s^*)|=M<\infty$. Draw the normal to ∂B at s^* and take t on it and inside B. If we set $|z^*-t|=\epsilon$, then by (7.21), neglecting higher order terms, $|z^*-\overline{S(t)}|=\epsilon$, $|t-\overline{S(t)}|=2\epsilon$. Furthermore, for $z\in\partial B$ in a neighborhood of z^* we have $|z-t|=\sqrt{(s-s_0)^2+\epsilon^2}=|z-\overline{S(t)}|$, neglecting higher order terms. Since $u(s)$ is continuous, given $\delta>0$, we can find $\sigma>0$ such that $|u(s)-u(s^*)|\leqq\delta$ for all $s^*-\sigma\leqq s\leqq s^*+\sigma$. Designate this arc by C_2 and the complementary arc on ∂B by C_1. Then, from (13.5), we have

$$(13.6) \quad | u(t) - u(z^*) |$$

$$\leqq \frac{1}{2\pi} \int_{C_1} + \int_{C_2} | u(z) - u(z^*) | \frac{| t - \overline{S(t)} |}{| z - t | | z - \overline{S(t)} |} ds.$$

This leads to

$$| u(t) - u(z^*) |$$

$$\leqq \frac{M\epsilon}{\pi} \int_{C_1} \frac{ds}{| z - t | | z - \overline{S(t)} |} + \frac{\delta}{2\pi} \int_{s^*-\sigma}^{s^*+\sigma} \frac{2\epsilon}{(s - s^*)^2 + \epsilon^2} ds$$

$$\leqq \frac{M\epsilon}{\pi} \int_{C_1} \frac{ds}{| z - t | | z - \overline{S(t)} |} + \frac{2\delta}{\pi} \arctan \frac{\sigma}{\epsilon}.$$

Now, as $t \to z^*$, then $\epsilon \to 0$. For $z \in C_1$, $| z - t | | z - \overline{S(t)} |$ clearly has a positive lower bound η so that

$$(13.7) \quad | u(t) - u(z^*) | \leqq \frac{M\epsilon S}{\eta} + \frac{2\delta}{\pi} \arctan \frac{\sigma}{\epsilon}.$$

From (13.7), $\limsup_{t \to z^*} | u(t) - u(z^*) | \leqq \delta$, and since δ is arbitrarily small, we have $\lim_{t \to z^*} u(t) = u(z^*)$. This conclusion is independent of whether $u(t)$ is or is not continuable throughout B.

We have discussed the case in which $S(z)$ is meromorphic in B. It is also possible to modify the method of images so that it is applicable to a contour whose Schwarz Function is an algebraic function of a certain type. However, we shall content ourselves with a reference to the literature (see p. 215).

CHAPTER 14

SCHWARZ FUNCTIONS OF
SPECIFIED TYPE

In view of the fact that certain constructions require that $S(z)$ be meromorphic inside the curve, it is of considerable interest to characterize analytic curves with this property. We shall obtain a characterization in terms of the mapping function of the region bounded by the curve.

DEFINITION. Let B be a region of the complex plane. The class $\mathfrak{a}C$ $(=\mathfrak{a}C(B))$ will designate the functions f which are regular in B and continuous in $B + \partial B$.

DEFINITION. A linear functional L, defined on functions regular in B, is said to be of class $\mathfrak{D} = \mathfrak{D}(B)$ if it can be expressed in the form

$$(14.1) \qquad L(f) = \sum_{n=1}^{N} \sum_{k=0}^{n_k} a_{nk} f^{(k)}(z_n),$$

where $z_1, \cdots, z_N$ are distinct points in (the interior of) B and where a_{nk} are constants independent of f.

The class $\mathfrak{D}$ may be described as *the point differential functionals of finite order*.

As preparation for the next result, we quote a theorem of J. L. Walsh, which will be used in the proof.

THEOREM. *Let B be a bounded, simply connected region*

whose boundary C is an analytic Jordan curve. Let $t(z)$ be continuous on C. Then

$$(14.2) \qquad \int_C t(z)f(z)\,dz = 0$$

for every $f \in \mathfrak{a}C(B)$ if and only if there exists a function $g \in \mathfrak{a}C(B)$ which coincides with t on C.

We can now prove:

THEOREM. *Let B be a bounded, simply connected region with analytic boundary ∂B. Let the Schwarz Function of ∂B be $S(z)$. Then, $S(z)$ is meromorphic in B if and only if there is an $L \in \mathfrak{D}$ such that*

$$(14.3) \qquad \int_B \int f(z)\,dx\,dy = L(f)$$

for all functions $f \in \mathfrak{a}C(B)$.

Proof: Suppose, first, that $S(z)$ is meromorphic. We have by (11.33), for any $f \in \mathfrak{a}C(B)$, $\int_B \int f(z)\,dx\,dy = (1/2i) \int_{\partial B} S(z)f(z)\,dz$. Now, since S is regular on ∂B and meromorphic in B, $S(z)$ has only a finite number of poles, say at $z_1, \cdots, z_N \in B$. Hence,

$$\frac{1}{2i} \int_{\partial B} S(z)f(z)\,dz = \sum_{n=1}^{N} \frac{\pi}{2\pi i} \int_{C_n} S(z)f(z)\,dz,$$

where C_n is a circle of sufficiently small radius centered at z_n. Now, in C_n, $S(z)$ has an expansion of the form

$$S(z) = \zeta_n(z) + \frac{B_{1n}}{z - z_n} + \frac{B_{2n}}{(z - z_n)^2} + \cdots + \frac{B_{p_n n}}{(z - z_n)^{p_n}},$$

where $\zeta_n(z)$ is regular in the closure of C_n. Hence,

$$\frac{\pi}{2\pi i} \int_{C_n} S(z) f(z) dz = \pi \sum_{k=1}^{p_n} \frac{B_{kn} f^{(k-1)}(z_n)}{(k-1)!}$$

and therefore we can conclude that (14.3) is valid.

Conversely, suppose that (14.3) holds for some $L \in \mathfrak{D}$ given by (14.1), and for all $f \in \mathfrak{a}C(B)$. Let $R(z)$ be the rational function

$$R(z) = \frac{1}{\pi} \sum_{n=1}^{N} \sum_{k=1}^{n_k} \frac{k! a_{nk}}{(z - z_n)^{k+1}},$$

so that for all $f \in \mathfrak{a}C(B)$, one has $L(f) = (1/2i) \int_{\partial B} R(z) f(z) dz$. Therefore, for all $f \in \mathfrak{a}C(B)$, we have

$$\int_{\partial B} (S(z) - R(z)) f(z) dz = 0 \quad \text{for all} \quad f \in \mathfrak{a}C(B).$$

The function $S(z) - R(z)$ is certainly continuous on ∂B (in fact it is regular in an annulus-like strip containing ∂B). Hence, by Walsh's Theorem, there exists a function $\zeta(z) \in \mathfrak{a}C(B)$ such that $S(z) - R(z) = \zeta(z)$, $z \in \partial B$. The function $\zeta(z)$ therefore takes analytic values on ∂B, and hence it must be analytic in $B + \partial B$. It therefore has a common region of analyticity with $S(z) - R(z)$. By the Uniqueness Theorem, it follows that $S(z) - R(z) = \zeta(z)$ throughout B, so that $S(z) = R(z) + \zeta(z)$ and is therefore meromorphic in B.

At this stage of the analysis it is convenient (though not necessary) to work with the Hilbert space $L^2(B)$ of functions that are single-valued and analytic in B and are such that $\iint_B |f(z)|^2 dx dy < \infty$. We give a brief outline of the salient features of $L^2(B)$. The inner product is given by $(f, g) = \iint_B f(z) \overline{g(z)} dx dy$. Let B be a bounded simply connected region with boundary ∂B. Assume that the complement of $B + \partial B$ is a single region whose boundary is precisely ∂B. Then the set of powers $1, z, z^2, \cdots$ is complete in $L^2(B)$ and hence for such a region, there is a complete orthonormal set of polynomials $\{p_n{}^*(z)\}$. If B is a simply connected region in the z-plane that can be mapped 1-1 conformally onto $|w| \leqq 1$ by means of $w = M(z)$, then the functions $\phi_n{}^*(z) = \sqrt{(n+1)/\pi}(M(z))^n M'(z)$, $n = 0, 1, \cdots$ constitute a complete orthonormal set for $L^2(B)$.

If $\{\phi_n{}^*(z)\}$ is any complete orthonormal system for $L^2(B)$, the bilinear expansion $\sum_{n=0}^{\infty} \phi_n{}^*(z)\overline{\phi_n{}^*(w)}$ converges to the *Bergman reproducing kernel function* $K(z, \bar{w})$. The convergence is absolute and uniform for (z, w) in compact subsets of $B \times B$. This kernel has the characteristic property of reproducing functions of $L^2(B)$ according to the formula $f(w) = (f(z), K(z, \bar{w}))$. If z_0 is a point in the interior of B then for each n the linear functional $L_n(f) = f^{(n)}(z_0)$, $n = 0, 1, \cdots$ is bounded over $L^2(B)$. If C is a rectifiable arc lying in the interior (endpoints included) of B and if $\int_C |w(z)| \, |dz| < \infty$, then the linear functional $L(f) = \int_C w(z)f(z)dz$ is bounded over $L^2(B)$. If L is any bounded linear functional over $L^2(B)$, its *representer* $r(z)$ is given by

$$(14.4) \qquad r(z) = \overline{L_w K(\bar{z}, w)}.$$

This means that

$$(14.5) \quad L(f) = (f, r) = \int_B \int f(z)\overline{r(z)}dxdy, \quad f \in L^2(B).$$

If B is a simply connected region and if $w = M(z)$, $M(0) = 0$, performs a 1-1 conformal map of B onto the unit circle $|w| \leqq 1$, then the Bergman kernel function can be expressed in terms of $M(z)$ as follows:

$$K(\bar{z}, w) = \frac{1}{\pi} \frac{\overline{M'(z)} M'(w)}{(1 - \overline{M(z)} M(w))^2}.$$

If $L(f) = \iint_B f(z)dxdy$, then $r(z) \equiv 1$ and we have

(14.6) $$\overline{L_\eta K(\bar{z}, \eta)} \equiv 1.$$

This is the functional equation for functionals whose representer is identically one. If we assume that B is simply connected, then

(14.7) $$1 = \frac{1}{\pi} L_\eta \left(\overline{\frac{\overline{M'(z)} M'(\eta)}{(1 - \overline{M(z)} M(\eta))^2}} \right),$$

or

(14.8) $$1 \equiv \frac{1}{\pi} \sum_{n=0}^{\infty} (n+1)(M(z))^n M'(z) \overline{L((M(\eta))^n M'(\eta))},$$

where the convergence is absolute and uniform on every compact subregion of B. Thus,

(14.9) $$1 = \frac{1}{\pi} \frac{d}{dz} \sum_{n=0}^{\infty} (M(z))^{n+1} \overline{L(M^n(\eta) M'(\eta))}.$$

Integrating from $z = 0$ to $z = z$ in B,

(14.10) $$\pi z = M(z) \sum_{n=0}^{\infty} (M(z))^n \overline{L(M^n(\eta) M'(\eta))}.$$

From this it follows that

(14.11) $$\bar{z} = \overline{M(z)} L_\eta \left(\frac{M'(\eta)}{1 - M(\eta) \overline{M(z)}} \right).$$

If $z = m(w), m(0) = 0$, designates the function inverse to

$w = M(z)$, (14.11) can be written as

$$(14.12) \qquad \pi \overline{m(w)} = \bar{w} L_\eta \left(\frac{M'(\eta)}{1 - \bar{w} M(\eta)} \right).$$

THEOREM. *Let B be a bounded simply connected region whose boundary ∂B is analytic and has Schwarz Function $S(z)$. Assume that $z = 0$ is contained in B and that $z = m(w)$, $m(0) = 0$ maps the unit circle $|w| \leqq 1$ one to one conformally onto B. Then $S(z)$ is meromorphic in B if and only if $m(w)$ is a rational function of w.*

Proof: Suppose that $m(w)$ is rational; we can write

$$m(w) = \frac{aw(1 - \beta_1 w) \cdots (1 - \beta_p w)}{(1 - \alpha_1 w) \cdots (1 - \alpha_n w)} = aw + \cdots,$$

where the α's differ from the β's. We have

$$\int_B \int f(z) \, dx \, dy = \frac{1}{2i} \int_{\partial B} \bar{z} f(z) \, dz$$

$$= \frac{1}{2i} \int_{|w|=1} \overline{m(w)} f(m(w)) m'(w) \, dw$$

$$= \frac{1}{2i} \int_{|w|=1} \bar{m} \left(\frac{1}{w} \right) f(m(w)) m'(w) \, dw.$$

Now,

$$\bar{m} \left(\frac{1}{w} \right) = \frac{\bar{a} w^{n-p-1} (w - \bar{\beta}_1) \cdots (w - \bar{\beta}_p)}{(w - \bar{\alpha}_1) \cdots (w - \bar{\alpha}_n)}.$$

If $n \geqq p + 1$, then $\bar{m}(1/w)$ has poles at $\bar{\alpha}_1, \cdots, \bar{\alpha}_n$ and no other singularities. If $n < p + 1$ then $\bar{m}(1/w)$ also has a pole of order $p + 1 - n$ at 0.

If we first assume that $n \geqq p + 1$ and that the points

α_i are distinct, then by the residue theorem,

$$\int_B \int f(z)\,dx\,dy$$

$$= \pi\bar{a} \sum_{k=1}^{m} f(m(\bar{\alpha}_k))\,m'(\bar{\alpha}_k)\, \frac{\bar{\alpha}_k{}^{n-p-1}(\bar{\alpha}_k - \bar{\beta}_1)\cdots(\bar{\alpha}_k - \bar{\beta}_p)}{\bar{P}'(\alpha_k)},$$

$$P(w) = (w - \alpha_1)\cdots(w - \alpha_n).$$

Thus, we have an identity of the form $\int_B \int f(z)\,dx\,dy = \sum_{k=1}^{n} c_k f(z_k)$, where c_k and the abscissas $z_k = m(\bar{\alpha}_k)$ are independent of f.

If the α_i are not distinct, then each point of higher multiplicity τ_k contributes a differential operator of order $\tau_k - 1$ evaluated at $\bar{\alpha}_k$.

If $n < p + 1$, then the point $\alpha = 0$ is a pole and hence $f(0)$ $(m(0) = 0)$ is present by itself if $n - p - 1 = -1$ or with its higher derivatives if $n - p - 1 < -1$.

In any case, therefore, $\int_B \int f(z)\,dx\,dy = L(f)$, $L \in \mathfrak{D}$, for all $f \in \mathcal{Q}C(B)$. By the last theorem, therefore, $S(z)$ must be meromorphic in B.

Conversely, suppose that $S(z)$ is meromorphic in B. Then by the last theorem, for $f \in \mathcal{Q}C(B)$, $\int_B \int f\,dx\,dy = L(f)$, with $L \in \mathfrak{D}$. Now by (14.12),

$$\overline{\pi m(w)} = \bar{w} L_\eta \left(\frac{M'(\eta)}{1 - \bar{w} M(\eta)} \right).$$

Since L is a point differential functional, it follows that $L_\eta((M'(\eta))/(1 - \bar{w}M(\eta)))$ and hence $\overline{m(w)}$ is a rational function of $\bar{w}$.

Examples. (a) $m(w) = w + aw^2$, $m'(w) = 1 + 2aw$, $m''(w) = 2a$; $\bar{m}(w) = w + \bar{a}w^2$ (limaçon).

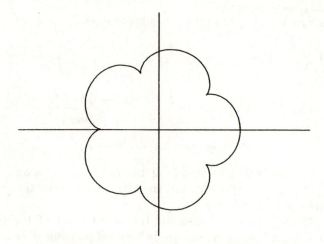

FIG. 14.1 : $z = w + \frac{1}{6}w^6$ (epitrochoid)

For $|a|$ sufficiently small ($|a| \leq \frac{1}{2}$), $m(w)$ is univalent in the unit circle and hence it maps onto a region B. From the example following (8.4′),

$$S(z) = \frac{1}{4z^2}(R+1)(2z + \bar{a}(R+1)),$$

where $\quad R = \sqrt{1 + 4az}.$

Therefore $S(z)$ has a pole of second order at $z = 0$ and no other singularities inside the limaçon. Now

$$\int_B \int f(z)\,dx\,dy = \pi \frac{1}{2\pi i} \int_{|w|=1} f(m(w))m'(w)\left[\frac{1}{w} + \frac{\bar{a}}{w^2}\right] dw$$

$$= \pi[(1 + 2|a|^2)f(0) + \bar{a}f'(0)].$$

Similar formulas can be worked out for $m(w) = w + aw^n$ (epitrochoid) (see Figs. 14.1–14.2).

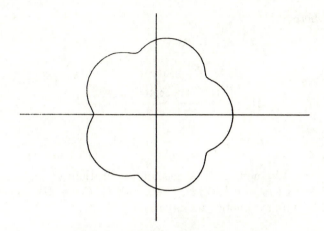

FIG. 14.2 : $z = w + \frac{1}{10}w^6$ (epitrochoid)

(b) Let

$$m(w) = \frac{w(1 - \beta^2 w^2)}{1 - \alpha^2 w^2},$$

$\beta \neq \pm\alpha$. Take α and β sufficiently small so that $m(w)$ is univalent in the circle $|w| \leqq 1$ and maps it onto a region B. Then

$$m'(w) = \frac{1 + \alpha^2 w^2 - 3\beta^2 w^2 + \alpha^2\beta^2 w^4}{(1 - \alpha^2 w^2)^2}$$

$$\bar{m}\left(\frac{1}{w}\right) = \frac{1}{w} \cdot \frac{(w^2 - \bar{\beta}^2)}{(w^2 - \bar{\alpha}^2)}.$$

Thus, $\int_B \int f(z)\,dx\,dy = Af(z^*) + Bf(0) + Af(-z^*),$

where

$$z^* = \frac{\bar{\alpha}(1 - \beta^2 \bar{\alpha}^2)}{1 - |\alpha|^4},$$

$$A = \frac{1 + |\alpha|^4 - 3\beta^2 \bar{\alpha}^2 + \beta^2 \bar{\alpha}^2 |\alpha|^4}{1 - |\alpha|^4},$$

$$B = (\bar{\beta}/\bar{\alpha})^2.$$

(c) We now give a second application of (14.12). Select $L(f) = \int_{-1}^{+1} f(\eta)\,d\eta$. Then, with $M(1) = -M(-1) = \alpha > 0$ for symmetry, we have from (14.12),

$$\pi \overline{m(w)} = \bar{w}\frac{1}{\bar{w}} \log\left(\frac{1 + \bar{w}\alpha}{1 - \bar{w}\alpha}\right) = \log\left(\frac{1 + \bar{w}\alpha}{1 - \bar{w}\alpha}\right).$$

Therefore

$$(14.13) \qquad z = \frac{1}{\pi} \log\left(\frac{1 + w\alpha}{1 - w\alpha}\right).$$

Inserting $z = 1$, $w = \alpha$ in (14.13) gives us $e^{\pi} = (1 + \alpha^2)/(1 - \alpha^2)$. Therefore,

$$(14.14) \qquad \alpha = \sqrt{\frac{1 - e^{-\pi}}{1 + e^{-\pi}}} \approx .958 < 1.$$

Observe that with $0 < \alpha \approx .958 < 1$, as w traces the unit circle, $\zeta = (1 - \alpha w)/(1 + \alpha w)$ traces a circle that lies in $\mathrm{Re}\,\zeta > 0$. Its center is at $((1 + \alpha^2)/(1 - \alpha^2), 0)$ and has radius $2\alpha/(1 - \alpha^2)$. The image in the z-plane of $|w| = 1$ under (14.13) is therefore schlicht. Since $|m(\pm 1)| = \alpha < 1$ and $m(z)$ is real for z real, the segment

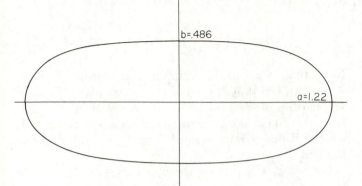

FIG. 14.3

$[-1, 1]$ is interior to this image. The map (14.13) therefore defines a simply connected region B.

The region B is an ellipse-like figure having a semi-major
axis $a = 1/\pi \log ((1 + \alpha)/(1 - \alpha)) \approx 1.22$ and a semi-
minor axis $b = 1/\pi \arctan (2\alpha/(1 - \alpha^2)) \approx .486$ (see Fig.
14.3).

The region B is such that

$$(14.15) \qquad \int_B \int f(z)\, dx dy = \int_{-1}^{+1} f(x)\, dx$$

for all $f \in \alpha C(B)$.

This identity can also be derived through the methods
of Chapter 11.

We have $m(w) = (1/\pi) \log ((1 + \alpha w)/(1 - \alpha w))$,

$M(z) = 1/\alpha((1 - e^{\pi z})/(1 + e^{\pi z}))$. Now from (8.4),

$$(14.16) \quad S(z) = \bar{m}\left(\frac{1}{M(z)}\right) = \frac{1}{\pi}\log\left(\frac{1 - e^{\pi}e^{\pi z}}{e^{\pi} - e^{\pi z}}\right).$$

Note that $S(z)$ has logarithmic singularities at $z = \pm 1 \pm 2ki$, $k = 0, 1, 2, \cdots$ and at no other place. The only singularities of $S(z)$ within B are therefore at $z = \pm 1$. By making a cut along the real axis from $z = 1$ to $z = -1$, we can define a single-valued branch of $S(z)$ inside B thus cut.

For $-1 \leqq x \leqq 1$ along the upper edge of the cut, we take

$$(14.17) \quad S_u(z) = S_{\text{upper}} = \frac{1}{\pi}\log\left(\frac{e^{\pi x}e^{\pi} - 1}{e^{\pi} - e^{\pi x}}\right) - i.$$

Along the lower edge of the cut we take

$$(14.18) \quad S_l(z) = S_{\text{lower}} = \frac{1}{\pi}\log\left(\frac{e^{\pi x}e^{\pi} - 1}{e^{\pi} - e^{\pi x}}\right) + i.$$

Now, $\int_B \int f(z)\,dx\,dy = (1/2i)\int_{\partial B} S(z)f(z)\,dz$. We now replace ∂B by a circuit consisting of $-(1 - \epsilon) \leqq x \leqq 1 - \epsilon$ augmented by two circles of radius ϵ centered at $x = 1$ and $x = -1$. We obtain

$$(14.19) \quad \iint_B f(z)\,dx\,dy$$

$$= \frac{1}{2i}\int_{-1}^{+1}(S_l(x)f(x) - S_u(x)f(x))\,dx.$$

The limiting process is valid since $\lim_{\epsilon \to 0}\epsilon\log\epsilon = 0$. Hence, from (14.17) and (14.18), $\iint_B f(z)\,dx\,dy = \int_{-1}^{+1}f(x)\,dx$.

The higher complex moments now follow. From (11.32),

$$\int_B \int \bar{z}^p f(z)\,dx dy$$

$$= \frac{1}{2i(p+1)} \int_{\partial B} S^{p+1}(z) f(z)\,dz$$

(14.20)

$$= \frac{1}{2i(p+1)} \int_{-1}^{+1} (S_l{}^{p+1}(x) - S_u{}^{p+1}(x)) f(x)\,dx$$

$$= \frac{1}{2i(p+1)} \int_{-1}^{+1} \zeta_p(x) f(x)\,dx,$$

where

(14.21) $\quad \zeta_p(x) = \left(\dfrac{1}{\pi} \log\left(\dfrac{e^{\pi x} e^{\pi} - 1}{e^{\pi} - e^{\pi x}} \right) + i \right)^{p+1}$

$$- \left(\frac{1}{\pi} \log\left(\frac{e^{\pi x} e^{\pi} - 1}{e^{\pi} - e^{\pi x}} \right) - i \right)^{p+1}$$

$$= 2i \operatorname{Im} \left(\frac{1}{\pi} \log\left(\frac{e^{\pi x} e^{\pi} - 1}{e^{\pi} - e^{\pi x}} \right) + i \right)^{p+1}.$$

As particular examples, for $p = 1$, we obtain

(14.22) $\quad \displaystyle\int_B \int \bar{z} f(z)\,dx dy = \frac{1}{\pi} \int_{-1}^{+1} \log\left(\frac{e^{\pi x} e^{\pi} - 1}{e^{\pi} - e^{\pi x}} \right) f(x)\,dx.$

For $p = 2$, we have

$$(14.23) \quad \iint_B \bar{z}^2 f(z)\, dx\, dy = \frac{1}{\pi^2} \int_{-1}^{+1} \log^2\left(\frac{e^{\pi x} e^{\pi} - 1}{e^{\pi} - e^{\pi x}}\right) f(x)\, dx$$

$$- \frac{1}{3} \int_{-1}^{+1} f(x)\, dx.$$

Let us generalize (14.15) by asking for a region B for which

$$(14.24) \quad \int_B \int f(z)\, dx\, dy = \int_{-1}^{+1} f(x) h(x)\, dx,$$

where $h(x)$ is a fixed weight function. In (14.12) take $L_\eta(f) = \int_{-1}^{+1} f(\eta) h(\eta)\, d\eta$. Then, assuming that $m(w)$ is real on the real axis, (14.12) reduces to the integral equation

$$(14.25) \quad m(w) = \frac{w}{\pi} \int_{m^{-1}(-1)}^{m^{-1}(1)} \frac{h(m(u))\, du}{1 - wu}.$$

Assuming that $h(x) \geqq 0$ on $(-1, 1)$, $h(x) = h(-x)$, and $\int_{-1}^{+1} h(x)\, dx > 0$, F. Stenger has shown that there is a mapping function $m(u)$ which satisfies (14.25) and which maps the unit circle onto a schlicht region B containing $(-1, 1)$.

When the double integral $\int_B \int f(z)\, dx\, dy$ has an alternate representation as $L(f)$, then the singularity structure of the Schwarz Function of ∂B coincides with that of the function $\psi(z) = L_x(1/(z - x))$. We shall develop this idea for the case $L(f) = \int_a^b g(x) f(x)\, dx$, $-\infty < a < b < \infty$. Here $\psi(z) = \int_a^b g(x)/(z - x)\, dx$. It has been found convenient to take g from the class of functions that are continuous in $a < x < b$, satisfy a Lipshitz condition there, but possibly have integrable singularities at the endpoints a and b. More precisely, we assume

(1) For every closed subinterval $\overline{\mathfrak{L}}$ of $a < x < b$, $g(x)$ satisfies a Lipshitz condition, i.e. $|\, g(x_1) - g(x_2)\,| \leqq A\,|\,x_1 - x_2\,|^\mu$, for some $A > 0$, $\mu > 0$, and for all x_1, $x_2 \in \overline{\mathfrak{L}}$.

(2) In the neighborhood of a or b, $g(x)$ is of the form

$$g(x) = \frac{g^*(x)}{(x - c)^\alpha} \quad \text{for some} \quad 0 \leqq \alpha < 1,$$

where c is either a or b and where $g^*(x)$ satisfies the Lipshitz condition of (1) on $a \leqq x \leqq b$.

We shall designate the class of functions satisfying (1) and (2) as H^*. It is well known that if $g \in H^*$, the Cauchy integral

$$(14.26) \qquad \psi(z) = \int_a^b \frac{g(x)\,dx}{z - x}$$

is analytic in the whole z plane with the segment $a \leqq x \leqq b$ deleted. Furthermore, this integral transform may be inverted by the *Plemelj-Stieltjes formula*:

$$(14.27) \quad \psi(x - i0) - \psi(x + i0) = 2\pi i g(x), \quad a < x < b.$$

THEOREM. *Let B be a bounded, simply-connected (open) region with an analytic boundary ∂B. Assume that the segment $[a, b]$ is in B. Let $S(z)$ be the Schwarz Function of ∂B, let $g \in H^*$, and let $\psi(z)$ be given by (14.26). Then $S(z)$ is of the form*

$$(14.28) \qquad S(z) = \psi(z) + \phi(z),$$

with $\phi(z) \in \mathfrak{a}C(B)$, *if and only if*

$$(14.29) \qquad \int_B \int f(z)\,dxdy = \pi \int_a^b g(x)f(x)\,dx$$

for all $f \in \mathfrak{a}C(B)$.

Proof: Assume that (14.28) holds. Then for all $f \in \mathfrak{a}C(B)$, we have

$$\int_B \int f(z)\,dxdy = \frac{1}{2i}\int_{\partial B} S(z)f(z)\,dz$$

$$= \frac{1}{2i}\int_{\partial B} (\psi(z) + \phi(z))f(z)\,dz$$

$$= \frac{1}{2i}\int_{\partial B} \psi(z)f(z)\,dz$$

$$= \frac{1}{2i}\int_{\partial B} \left(\int_a^b \frac{g(x)\,dx}{z - x}\right)f(z)\,dz$$

$$= \pi \int_a^b \frac{g(x)}{2\pi i}\int_{\partial B} \frac{f(z)}{z - x}\,dx$$

$$= \pi \int_a^b g(x)f(x)\,dx,$$

the interchange of the integrals being justifiable under the assumptions.

Conversely, suppose that (14.29) holds for all $f \in \mathfrak{a}C(B)$. Then, by the above line of reasoning, $(1/2i)\int_{\partial B} S(z)f(z)\,dz = (1/2i)\int_{\partial B} \psi(z)f(z)\,dz$, so that $\int_{\partial B} (S(z) - \psi(z))f(z)\,dz = 0$ for all $f(z) \in \mathfrak{a}C(B)$. By Walsh's theorem, (14.28) follows.

It is interesting to note the connection between these identities and the Plemelj formula (14.27).

Assume that for $a < x < b$, $\lim_{y \to 0^+} \psi(x + iy) = \psi_u(x)$ and $\lim_{y \to 0^+} \psi(x - iy) = \psi_l(x)$. Then, by (14.28), for all $a < x < b$, $\lim_{y \to 0^+} S(x + iy)$ exists and equals some $S_u(x)$ and $\lim_{y \to 0^+} S(x - iy)$ exists and equals some $S_l(x)$. By (14.27), $S_l(x) - S_u(x) = 2\pi i g(x)$. Now,

$$\int_B \int f(z) \, dx \, dy = \frac{1}{2i} \int_{\partial B} S(z) f(z) \, dz.$$

By "shrinking" the contour of integration ∂B to the segment $[a, b]$, traversed twice (assuming proper justification), we obtain

$$\frac{1}{2i} \int_{\partial B} S(z) f(z) \, dz = \frac{1}{2i} \int_{-1}^{+1} S_l(x) f(x) \, dx - \frac{1}{2i} \int_{-1}^{+1} S_u(x) f(x) \, dx$$

$$= \pi \int_{-1}^{+1} g(x) f(x) \, dx$$

so that we recover (14.29). See also (11.48).

We shall now point out an interesting consequence of the identity (14.15).

LEMMA. *Let B be a bounded region in the complex plane. Then we can find a sequence of discs C_n: $|z - z_n| < r_n$, $n = 1, 2, \cdots$, such that (1) each C_n is in B, (2) discs are non-overlapping, and (3) $\sum_{n=1}^{\infty}$ area $(C_n) = \sum_{n=1}^{\infty} \pi r_n^2 =$ area (B).*

Such a system of circles is called a *complete packing* of B. A proof of this lemma is easily given.

LEMMA. *Let C designate the circle $|z - z_0| < r$. If $f \in \mathfrak{a}C(C)$ then $\int_C \int f(z) \, dx \, dy = \pi r^2 f(z_0)$.*

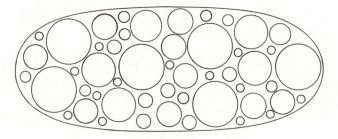

FIG. 14.4

This is a form of the mean value theorem for analytic functions and is a simple instance of the identities following (14.3).

THEOREM. *Let B designate the region in* (14.15). *Let $C_n: |z - z_n| < r_n$ be a complete packing of B by discs.* (See Fig. 14.4). *Then if $f(z) \in \text{α}C(B)$, we have*

$$(14.30) \qquad \int_{-1}^{1} f(x)\,dx = \pi \sum_{k=1}^{\infty} r_k^2 f(z_k).$$

A formula of the type $\int_a^b f(x)\,dx = \sum_{k=1}^{\infty} w_k f(x_k)$ with the x_k *distinct* is known as a *complete quadrature*. The above theorem shows how to obtain complete quadratures (with complex abscissas) for analytic functions. It is known that complete quadratures do not exist for the class $C[a, b]$ of real functions that are continuous on $[a, b]$. S. Haber has shown that they exist for the class $\text{Lip}^{\alpha}[a, b]$, $\alpha > 0$. Haber's complete quadratures are not absolutely convergent.

The formula (14.30) must be regarded as one of the curiosities of quadrature theory, rather than as a practical means of numerical integration. It is known that for complete packings $\sum_{n=1}^{\infty} r_n = \infty$, so that the convergence of (14.30), though absolute, is quite slow.

We have worked out a number of specific formulas which give alternate functional representations of double integrals of analytic functions: $\iint_B f(z)\,dx\,dy$. These include (11.31), (11.34), (11.40), (14.3), (14.15). We have also derived a functional equation (14.6) for the reverse problem: given a functional L, find a region B such that $\int_B \int f(z)\,dx\,dy = L(f)$ for all f analytic in B. We can call such a B (if it exists) *a characteristic region* for the functional L. It provides an easy way of visualizing L on a common geometrical basis. Many of the functionals met in numerical analysis are error functionals corresponding to various rules of interpolation, approximate integration, etc. These error functionals often have characteristic regions, and the study of these regions and their comparison for different rules can lead to interesting conclusions.

Consider, for example, the problem of approximate integration $\int_{-1}^{+1} f(x)\,dx \approx \sum_{k=1}^{m} c_k f(x_k)$. The functional $L(f) = \int_{-1}^{1} f(x)\,dx$ has the characteristic region B of (14.15). As we have seen, approximating functionals of the form $L_A(f) = \sum_{k=1}^{m} c_k f(x_k)$ and their characteristic regions can be constructed through the rational map

$$m(w) = \frac{aw(1 - \beta_1 w) \cdots (1 - \beta_p w)}{(1 - \alpha_1 w) \cdots (1 - \alpha_n w)},$$

$n \geqq p + 1$, α_i distinct. If the $c_i \geqq 0$, characteristic regions can also be constructed in an *ad hoc* manner using

$$\iint_{|z - z_0| \leqq r} f(z)\,dx\,dy = \pi r^2 f(z_0),$$

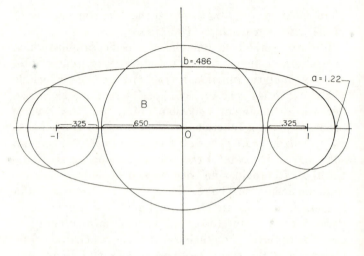

FIG. 14.5

as was done with the packings. By subtracting the two regions, we obtain a characteristic region for the error functional. Figure 14.5 shows this worked out for the error in Simpson's Rule $R(f) = \int_{-1}^{+1} f(x)\,dx - \frac{1}{3}(f(-1) + 4f(0) + f(1))$. The oval is approximated by the three circles. The difference region D is interpreted as the characteristic function of the former minus that of the latter, i.e., it consists of the oval with the value $+1$ attached together with the three circles with the value -1 attached. With this interpretation, $R(f) = \int_D \int f(z)\,dx\,dy$. Since Simpson's Rule integrates quadratic polynomials exactly, one has the identities $\int_D \int dx\,dy = \int_D \int z\,dx\,dy = \int_D \int z^2\,dx\,dy = 0$.

CHAPTER 15

SCHWARZ FUNCTIONS
AND ITERATION

The object of iteration theory is to study the properties of the set of successive transformations $\tau^n(P)$, $n = 0, 1, \cdots$ of the point P. Among the many questions of interest are:

(a) What are the invariant subsets of τ? What are the fixed points of τ? What are the periodic points (cycles) of τ? i.e., what are the points $P_1, P_2, \cdots, P_k$ such that $\tau(P_i) = P_{i+1}$ where the subscripts are taken mod k? What are the invariant curves?

(b) What are the convergence properties of $\tau^n(P)$? What are the limit sets?

(c) What are the *ergodic* properties of τ? i.e., what can be said about the averages of $\tau(P), \cdots, \tau^n(P)$?

(d) How can iteration be used to solve equations expeditiously?

For transformations τ with even the simplest analytic structure, the answers to these equations may be enormously complicated and there may be no general analytic techniques available for resolving them. Some recent investigators have used the computer, displaying the successive iterates on a computer scope, to gain insight into the nature of τ^n.

In this chapter, we shall deal only with the simplest transformations τ in two dimensions and limit ourselves to those questions in which the Schwarz Function plays a role.

15.1. Fixed points—generalities. Let $\tau:R^n \to R^n$ be a mapping from real n-dimensional space into itself.

The point set $\mathfrak{R}$ is *invariant*[*] under τ if and only if $\tau(\mathfrak{R}) = \mathfrak{R}$. P is a *fixed point* (or an *equilibrium point*) of τ if $\tau(P) = P$.

For the point P_0 consider the sequence

$$(15.1) \quad P_{n+1} = \tau(P_n) = \tau^{n+1}(P_0), \qquad n = 0, 1, \cdots.$$

A fixed point P is called *attractive* or *locally convergent* if there is a neighborhood $\mathfrak{N}$ of P such that whenever $P_0 \in \mathfrak{N}$, the successive iterates (15.1) may be formed and $\lim_{n\to\infty} P_n = P$.

There are various definitions of stability and we shall use the following one: τ is said to be *stable* at a fixed point P if every neighborhood $\mathfrak{N}$ of P contains an invariant neighborhood.

Suppose that τ maps a region $\mathfrak{R}$ into itself. Then τ is a *contraction mapping* if there is a constant $k: 0 < k < 1$ such that for all $P, Q \in \mathfrak{R}, \| \tau(P) - \tau(Q) \| \leq k \| P - Q \|$. The notation $\| \; \|$ designates the Euclidean norm in R^n, or more generally, a distance function in a metric space.

The importance of contraction maps lies in the fact that successive iterations converge to a fixed point. The following formulation of the contraction mapping theorem is of sufficient generality.

THEOREM. *Let X be a complete metric space and let $\tau: X \to X$ be a contraction map of X into itself. Then* (1)

---

[*] In this essay, $\mathfrak{R}$ is usually an analytic arc while τ is analytic, so that it is sufficient to assume that $\mathfrak{R}$ and $\tau(\mathfrak{R})$ are both subarcs of an analytic arc.

τ *has a unique fixed point* x^* *in* X. (2) $\lim_{n\to\infty} x_n = x^*$ *where* $x_{n+1} = \tau(x_n)$, $n = 0, 1, \cdots$ *and where* $x_0 \in X$ *is arbitrary.* (3) $\| x_n - x^* \| < (k^n/(1 - k)) \| x_1 - x_0 \|$.

The following is a *sufficient* condition for an attractive point.

THEOREM. *Let the transformation*

$$(15.2) \quad \tau: x_i' = f_i(x_1, \cdots, x_n), \qquad i = 1, 2, \cdots, n,$$

be of class C^1 *in the neighborhood of a fixed point* P: $x_1^*, \cdots, x_n^*$. *Let the Jacobian matrix at* P *be*

$$J = \left(\frac{\partial f_j}{\partial x_k}\right)_{x_i = x_i^*},$$

and let λ^+ *be the largest among the moduli of the eigenvalues of* J. *Then* P *is an attractive fixed point of* τ *if*

$$(15.3) \qquad\qquad \lambda^+ < 1.$$

Consider what this means in the special case

$$(15.4) \quad \begin{cases} x' = f(x, y) \\ y' = g(x, y) \end{cases} \text{ or } z' = x' + iy' = f + ig = F(z, \bar{z}),$$

where f and g are analytic functions. We have

$$(15.5) \qquad \lambda = \tfrac{1}{2}(T \pm \sqrt{T^2 - 4|J|}),$$

where $T = f_x + g_y = \text{trace}(J)$, $|J| = \det J$. In terms of conjugate coordinates, we have from (11.14)

$$(15.6) \quad T = F_z + \bar{F}_z, \qquad |J| = \begin{vmatrix} F_z & F_{\bar{z}} \\ \bar{F}_{\bar{z}} & \bar{F}_z \end{vmatrix}.$$

Further specialization to $F(z, \bar{z}) \equiv f(z)$, $\bar{F}(z, \bar{z}) \equiv \bar{f}(z)$, yields $T = \overline{f'(z)} + f'(z)$ and $|J| = |f'(z)|^2$ so that $\lambda_1 = f'(z)$, $\lambda_2 = \overline{f'(z)}$. At a fixed point z^*, the condition $\lambda^+ < 1$ becomes $|f'(z^*)| < 1$.

Here is a complex variable proof for attraction in this case.

THEOREM. *Let $f(z)$ be analytic in a neighborhood of $z = 0$. Suppose that $f(0) = 0$ and $f'(0) = s$, $0 < |s| < 1$. Then $z = 0$ is an attractive point for iteration under f.*

Proof: Note first that the zeros of $f(z)$ must be isolated. Hence there is a $\rho_1 > 0$ such that $z = 0$ is the only zero of f in $|z| < \rho_1$. Since $f(z)/z = s + bz + cz^2 + \cdots$, it follows that given a $t: |s| < t < 1$, one can find a $\rho_2 > 0$ such that $|f(z)/z| < t$ for $|z| < \rho_2$. Select $\rho = \min (\rho_1, \rho_2)$. Take any z_0 with $0 < |z_0| < \rho$. Then

$$z_1 = f(z_0) \neq 0 \quad \text{and} \quad \left| \frac{z_1}{z_0} \right| = \left| \frac{f(z_0)}{z_0} \right| < t,$$

so that $0 < |z_1| < t|z_0| < t\rho < \rho$. Similarly, $z_2 = f(z_1)$, so that $|z_2/z_1| = |f(z_1)/z_1| < t$. Therefore $0 < |z_2| < t|z_1| < t^2\rho < \rho$. In general, $0 < |z_n| < t^n\rho < \rho$, so that the sequence $\{z_n\}$ consists of distinct points with $\lim_{n\to\infty} z_n = 0$. The convergence is uniform in closed subsets of $|z| < \rho$.

If $f(z)$ is defined in a region R which is closed, bounded and convex, it can be shown that a necessary and sufficient condition that $f(z)$ be a contraction map is that $|f'(z)| \leqq k < 1$ in R. However, the condition of contraction is of course not necessary to produce a unique fixed point. In addition to (15.3), one has *the Henrici Fixed Point Theorem*.

THEOREM. *Let R be the interior of a Jordan curve C. Let*

*$f(z)$ be regular and analytic in R and continuous in $R + C$.
Suppose that f maps $R + C$ into R. Then f has exactly one
fixed point and simple iteration $z_{n+1} = f(z_n)$ converges to it
starting from an arbitrary point $z_0 \in R + C$.*

Proof: Consider first the case where C is the unit circle.
The hypothesis implies that $\max_{|z| \leq 1} |f(z)| < 1$. Now s is
a fixed point if and only if it is a zero of $z - f(z)$. It follows
from Rouché's theorem (see, e.g., Ahlfors [**A1**] p. 124)
that $z - f(z)$ has as many zeros in $|z| \leq 1$ as does z, i.e.,
one.

If s designates the fixed point, the bilinear transformation

$$T(z) = \frac{z - s}{1 - z\bar{s}}$$

maps $|z| \leq 1$ onto itself, preserves the unit circle and
sends s to 0. (See p. 59.) Designate its inverse by T^{-1}
and consider the composite function

$$G = TfT^{-1}$$

which has 0 as a fixed point.

This function is continuous in $|z| \leq 1$, mapping the
disc onto a closed subset of $|z| < 1$. Hence, $0 \leq k =
\max_{|z| \leq 1} |G(z)| < 1$. Assume that $k > 0$; otherwise G,
and hence f, are constant and convergence occurs in one
step. The function G/k is 0 at 0 and $|G/k| \leq 1$ in $|z| \leq 1$.
By the Schwarz Lemma [see, e.g., Ahlfors [**A1**] p. 110]
$|G(z)/k| \leq |z|$ in $|z| \leq 1$ and hence $|G(z)| \leq k|z|$
there. Now, set $w_n = T(z_n)$, then

$$w_{n+1} = T(z_{n+1}) = T(f(z_n)) = TfT^{-1}(w_n) = G(w_n),$$

and it follows that

$$|w_{n+1}| \leqq k|w_n|$$

so that $|w_n| \leqq k^n |w_0|$. Therefore $w_n \to 0$ and hence $z_n = T^{-1}(w_n) \to T^{-1}(0) = s$.

If now C is a general Jordan curve, we employ the Riemann mapping theorem with the formulation of Osgood-Carathéodory which guarantees the existence of an analytic g that maps R conformally onto $|z| < 1$ and maps $R + C$ continuously and one to one onto $|z| \leqq 1$. [See, e.g., Ahlfors [A1] p. 92–98.]

The composite function $h = gfg^{-1}$ now satisfies the hypothesis for the theorem for the unit disc. If $w_n = g(z_n)$, then $w_{n+1} = h(w_n)$. The conclusion follows from the first part of the proof.

15.2. Iteration of Schwarz Functions. The equation $z = \overline{S(z)}$ for an analytic curve C would inevitably suggest iteration and the contraction mapping theorem to a numerical or a functional analyst. The numerical analyst works with the equation $x = f(x)$, where x is generally a real variable. However, there are differences to be noted. The map $z \to \overline{S(z)}$ cannot be a contraction map; if it were, we should have a unique solution, which is impossible. Furthermore, $|S'(z)| = 1$ along C. And, in point of fact, simple iteration $z_{n+1} = \overline{S(z_n)}$ is cyclic (for $\bar{S}S = I$) and hence divergent unless $z_1 \in C$. The anti-analytic nature of $\overline{S(z)}$ is another feature.

Following a geometric lead, for points z near to C, $\overline{S(z)}$ is nearly the geometric reflection of z in C. Hence, by the remark following (7.21), the point $\frac{1}{2}(z + \overline{S(z)})$ very nearly lies on C. This suggests that one study the iteration

$$(15.7) \qquad z_{n+1} = \tfrac{1}{2}(z_n + \overline{S(z_n)})$$

or, more generally,*

(15.8) $z_{n+1} = (1 - t)z_n + t\overline{S(z_n)}, \qquad 0 < t < 1.$

We shall prove, in fact, that for all points sufficiently close to C, the scheme (15.8) converges with at least linear rapidity to a point $z^* \in C$, while (15.7) converges at least quadratically.

For the iteration (15.8) we have

$$F(z, \bar{z}) = (1 - t)z + t\overline{S(z)}, \qquad |1 - 2t| < 1.$$

From (15.5) we have $\lambda = (1 - t) \pm t\,|\,S'(z)\,|$. Since $|\,S'(z)\,| = 1$ along C, $\lambda^+ = 1$. Thus, the iteration (15.8) provides an interesting example demonstrating that the condition $\lambda^+ < 1$ is not necessary for convergence in a neighborhood of a fixed point. (Of course, the limit of z_n depends upon the starting value z_0.)

Note that if C is the circle $|\,z\,| = R$, (15.7) reduces to

$$(15.9) \qquad z_{n+1} = \frac{1}{2}\left(z_n + \frac{R^2}{\bar{z}_n}\right)$$

which, in the real case, is the familiar square root algorithm. The iteration (15.8) is

$$(15.10) \qquad z_{n+1} = (1 - t)z_n + \frac{tR^2}{\bar{z}_n}\,.$$

The real form of these iterations is the Newton Method

$$z_{n+1} = z_n - \frac{f(z_n)}{f'(z_n)}$$

applied to the function $f(z) = (z^2 - R^2)^{1/2t}$.

---

* Numerical iterations of considerable complexity have been studied. See, e.g., Traub [T1], Rall [R1]. However, we shall confine ourselves to iterations which follow the "square root" pattern, as suggested by the functional equation itself.

Let us observe first of all that if C is an analytic arc and $z_0 \in C$, then we can find a sufficiently small neighborhood of z_0, $U = U(z_0)$, such that if $z \in U$ then $(1 - t)z + t\overline{S(z)} \in U$. Thus, if $z_n \in U$, $z_{n+1} \in U$ and all the iteration action takes place in U.

To show this, suppose that $z = f(t)$, $z_0 = f(t_0)$, $0 \leqq t \leqq 1$, $0 < t_0 < 1$, is the parametric representation of C. There is a disc $|t - t_0| \leqq \rho$ whose image V under f is schlicht. Now, by a well-known theorem of Study (see, e.g., Hille [H4], vol. II, p. 359) there is a subdisc $D\colon |t - t_0| \leqq \sigma < \rho$ whose image U under f is convex. Select $z \in U$; then $f^{-1}(z) \in D$. Hence $\overline{f^{-1}(z)} \in D$ so that $f(\overline{f^{-1}(z)}) = \overline{S(z)} \in U$. Since it is convex, $(1 - t)z + t\overline{S(z)} \in U$.

However, the proof which follows does not make use of this.

LEMMA 1. *Let C be a closed (i.e., C bounds a simply connected region) analytic curve with Schwarz Function $S(z)$. Let $S(z)$ have a single-valued and regular branch in a closed annulus-like region B that contains C in its interior. Let $z_0 \in C$ and set*

$$(15.11) \quad S(z) \equiv S(z_0) + S'(z_0)(z - z_0) + (z - z_0)^2 \epsilon(z; z_0).$$

Then, we can find a constant M such that

$$(15.12) \quad |\epsilon(z; z_0)| \leqq M \quad for\ all \quad z_0 \in C, z \in B.$$

Proof: We can find an annulus-like region B^* containing B in its interior in which $S(z)$ remains regular and single-valued. We may suppose further that B^* has smooth outer and inner boundary curves Γ_1 and Γ_2. By the complex form of Taylor's theorem with the exact remainder, we have for fixed z_0 and z,

$$(15.13) \qquad \epsilon(z; z_0) = \frac{1}{2\pi i} \int_G \frac{S(t)\,dt}{(t - z_0)^3(t - z)},$$

where G is a contour which contains z_0 and z in its interior and is itself contained in B^*. By a standard argument, we may replace G by $\Gamma_1 - \Gamma_2$ so that

$$(15.14) \quad \epsilon(z; z_0) = \frac{1}{2\pi i} \int_{\Gamma_1} \frac{S(t)\,dt}{(t - z_0)^3(t - z)}$$

$$- \frac{1}{2\pi i} \int_{\Gamma_2} \frac{S(t)\,dt}{(t - z_0)^3(t - z)}.$$

If $S_i = \max_{t \in \Gamma_i} | S(t) |$, $L_i = $ length of Γ_i, $\delta_i = $ min dist from Γ_i to C, $\Delta_i = $ min dist from Γ_i to B, then we have for all $z_0 \in C, z \in B$,

$$(15.15) \qquad | \epsilon(z; z_0) | \leqq \frac{1}{2\pi} \sum_{i=1}^{2} \frac{S_i L_i}{\delta_i^3 \Delta_i} \equiv M.$$

Let C be as before and for each sufficiently small δ, let $C_{\pm\delta}$ designate the two curves "parallel" to C and at a distance of δ from it. That is, each point z on C_δ is at a minimum distance of δ from C, $P(z)$ being the unique relevant orthogonal projection of z on C. C_δ contains C in its interior while $C_{-\delta}$ is contained in C. Let $D_\delta = \bigcup_{0 \leq \Delta \leq \delta} C_{\pm\Delta}$. D_δ is an annulus-like region containing C in its interior all of whose points lie at a distance $\leqq \delta$ from C.

Let t be a fixed complex number located in the circle $| t - \frac{1}{2} | < \frac{1}{2}$. Let $\lambda = 1 - 2t$. Then $| \lambda | < 1$.

With C and B fixed, as in Lemma 1, and variable σ, set

$\tau = |\lambda| + |t| \sigma M$. It is clear that σ can be selected so small that (1) D_σ is contained in the interior of B and (2) $0 < \tau < 1$.

LEMMA 2. *Let* $z_0 \in D_\sigma$; *set* $z_1 = (1 - t)z_0 + t\overline{S(z_0)}$, *and let* P *designate the projection explained above. Then,*

$$(15.16) \quad |z_0 - P(z_0)| \leqq \sigma, \qquad |z_1 - P(z_0)| \leqq \sigma\tau,$$

and $z_1 \in D_\sigma$.

Proof: Without loss of generality we may assume that z_0 is located on the imaginary axis and that $P(z_0) = 0$. That is, C passes through 0 and has 0 slope there. Therefore from (7.23) and (15.11) we may write

$$S(z) = z + z^2\epsilon(z; 0) \quad \text{for} \quad z \in B.$$

Now, by our assumption, $z_0 = i\rho$ with $|\rho| \leqq \sigma$ so that

$$\overline{S(z_0)} = -i\rho - \rho^2\overline{\epsilon(i\rho, 0)}.$$

Hence,

$$z_1 = (1 - t)i\rho + t(-i\rho - \rho^2\overline{\epsilon(i\rho, 0)})$$
$$= (1 - 2t)i\rho - t\rho^2\overline{\epsilon(i\rho, 0)}.$$

Therefore $|z_1 - P(z_0)| = |z_1| \leqq |\lambda|\rho + |t|\rho^2 M \leqq (|\lambda| + |t|\sigma M)\sigma = \tau\sigma < \sigma$. Now, $|z_1 - P(z_1)| \leqq |z_1 - P(z_0)| \leqq \sigma$ (since the projection of z_1 on C is $P(z_1)$) and hence $z_1 \in D_\sigma$.

THEOREM. *Let* $z_0 \in D_\sigma$. *Then the iteration*

$$(15.17) \quad z_{n+1} = (1 - t)z_n + t\overline{S(z_n)} \qquad n = 0, 1, 2, \cdots$$

can be carried out. The sequence of points $\{z_n\}$ *generated converges to a point* $z^* \in C$.

Proof: From Lemma 2, $z_1 \in D_\sigma$; hence $S(z_1)$ is defined and z_2 can be formed. We now iterate this. Applying Lemma 2 again, with z_1 in the role of z_0, we have, $|z_1 - P(z_1)| \leqq \sigma\tau$, $|z_2 - P(z_1)| \leqq \sigma\tau^2$, $|z_2 - P(z_2)| \leqq \sigma\tau^2$, $\cdots$, $|z_n - P(z_n)| \leqq \sigma\tau^n$, $|z_{n+1} - P(z_n)| \leqq \sigma\tau^{n+1}$. Therefore $|z_{n+1} - z_n| \leqq \sigma(1 + \tau)\tau^n$ so that $\sum_{i=0}^{\infty} |z_{i+1} - z_i| < \infty$. Therefore $\sum_{i=0}^{\infty} (z_{i+1} - z_i)$ converges. But $\sum_{i=0}^{n} (z_{i+1} - z_i) = z_{n+1} - z_0$ and hence z_{n+1} converges. Call its limit z^*. We have $z^* \in D_\sigma$, but allowing $n \to \infty$ in (15.8) we have by continuity, $z^* = (1 - t)z^* + t\overline{S(z^*)}$ or $z^* = \overline{S(z^*)}$ so that $z^* \in C$.

THEOREM. *The convergence of $\{z_n\}$ to z^* is at least linear in rapidity. If $t = \frac{1}{2}$, the convergence is at least quadratic.*

Proof: We have $z^* - z_n = (z_{n+1} - z_n) + (z_{n+2} - z_{n+1}) + \cdots$ so that

$$|z^* - z_n| \leqq |z_{n+1} - z_n| + |z_{n+2} - z_{n+1}| + \cdots$$

$$\leqq \sigma(1 + \tau)(\tau^n + \tau^{n+1} + \cdots) = \sigma\frac{1 + \tau}{1 - \tau}\tau^n.$$

If $t = \frac{1}{2}$, then $\lambda = 0$. Referring to the proof of the previous theorem, we can assert that $|z_0 - P(z_0)| \leqq \sigma$ and $|z_1 - P(z_0)| \leqq (M/2)\sigma^2 = \mu\sigma^2$ with $\mu = M/2$. Note that $\mu\sigma < 1$. Therefore $|z_1 - P(z_1)| \leqq \mu\sigma^2$ and $|z_2 - P(z_2)| \leqq \mu(\mu\sigma^2)^2 = \mu^3\sigma^4$. In general, $|z_n - P(z_n)| \leqq (1/\mu)(\mu\sigma)^{2^n}$ so that $|z_{n+1} - z_n| \leqq 1/\mu((\mu\sigma)^{2^n} + (\mu\sigma)^{2^{n+1}})$. Then,

$$|z^* - z_n| \leqq (2/\mu)[(\mu\sigma)^{2^n} + (\mu\sigma)^{2^{n+1}} + (\mu\sigma)^{2^{n+2}} + \cdots]$$

$$\leqq \frac{2}{\mu}\frac{(\mu\sigma)^{2^n}}{1 - (\mu\sigma)^{2^n}}.$$

15.3. Invariance of analytic curves.   A transformation τ of the plane can be specified by

$$(15.18) \qquad \tau : \begin{cases} u = f(x, y) \\ v = g(x, y). \end{cases}$$

We shall suppose that f and g are analytic functions of two variables x and y in an appropriate domain. We can write (15.18) in terms of conjugate variables as follows:

$$(15.19) \quad w = u + iv$$

$$= f\left(\frac{z + \bar{z}}{2}, \frac{z - \bar{z}}{2i}\right) + ig\left(\frac{z + \bar{z}}{2}, \frac{z - \bar{z}}{2i}\right)$$

or

$$(15.20) \qquad \tau : w = F(z, \bar{z}).$$

It is convenient to assume that $z = 0$ is a fixed point of τ so that $F(0, 0) = 0$. In particular, the linear transformation

$$(15.21) \quad \begin{cases} u = ax + by \\ v = cx + dy \end{cases}, \qquad \begin{vmatrix} a & b \\ c & d \end{vmatrix} \neq 0, \quad a, b, c, d \text{ real}$$

takes the form

$$(15.22) \qquad w = Az + B\bar{z} \equiv F(z, \bar{z})$$

with

$$(15.23) \quad \begin{cases} A = \frac{1}{2}((a + d) + i(c - b)) \\ B = \frac{1}{2}((a - d) + i(c + b)). \end{cases}$$

This can be written in matrix form as

$$(15.24) \quad \begin{pmatrix} w \\ \bar{w} \end{pmatrix} = P \begin{pmatrix} z \\ \bar{z} \end{pmatrix}, \qquad P = \begin{pmatrix} A & B \\ \bar{B} & \bar{A} \end{pmatrix},$$

so that

$$P^{-1} = 1/|J| \begin{pmatrix} \bar{A} & -B \\ -\bar{B} & A \end{pmatrix},$$

where $|J| = |A|^2 - |B|^2$. Furthermore, P and $Q = \begin{pmatrix} a & b \\ c & d \end{pmatrix}$ are similar and hence their eigenvalues coincide (see (9.19)).

Suppose that the analytic curve S with Schwarz Function $S(z)$ is invariant under τ. If z is on S, then $\bar{z} = S(z)$. But w is also on S, so that $\bar{w} = S(w)$. Therefore $\bar{w} = S(F(z, \bar{z}))$. But $\bar{w} = \overline{\tau(z)} = \overline{F(z, \bar{z})} = \bar{F}(\bar{z}, z)$. Hence

$$(15.25) \qquad S(F(z, S(z))) = \bar{F}(S(z), z).$$

Suppose, conversely, that the Schwarz Function $S(z)$ of an analytic curve S satisfies (15.25). Then it follows that if $z \in S$, $w \in S$. Thus, (15.25) *is the functional equation for invariance of an analytic curve.*

As a particular instance, under the linear transformation (15.22), the Schwarz Function $S(z)$ of an invariant analytic curve must satisfy

$$(15.26) \qquad S(Az + BS(z)) = \bar{A}S(z) + \bar{B}z.$$

Let us look for the invariant subspaces of τ (i.e., straight lines through the origin) with Schwarz Function $S(z) = \omega z$, $|\omega| = 1$. Equation (15.26) reduces to

$$(15.27) \qquad B\omega^2 + (A - \bar{A})\omega - \bar{B} = 0.$$

The roots are

$$\omega_1 = \frac{\rho + \Delta^{1/2}}{2B}, \qquad \omega_2 = \frac{\rho - \Delta^{1/2}}{2B},$$

$$\rho = \bar{A} - A = -2i \operatorname{Im} A, \qquad \Delta = \rho^2 + 4|B|^2 =$$

$4(|B|^2 - (\text{Im } A)^2)$. If $\Delta > 0$, then the ω_i are distinct and $|\omega_i| = 1$. The lines $\bar{z} = \omega_1 z$ and $\bar{z} = \omega_2 z$ are invariant.

On the basis of (15.26), it is easily shown that the circles $S(z) = r^2/z$ are invariant if and only if $|A| = 1$, $B = 0$ or $A = 0$, $|B| = 1$ (rotations or reflections) whereas the curve $|A| \neq 1$, $B = 0$ or $A = 0$, $|B| \neq 1$ has the spirals of Bernoulli $S(z) = z^\omega$ ($\omega = (\log \bar{A}/\log A)$ in the former case) as invariant curves.

If $\Delta > 0$, the matrix Q or P has eigenvalues λ_1, λ_2 that are *real and distinct*. Therefore Q can be diagonalized by a real T:

$$T^{-1}QT = \begin{pmatrix} \lambda_1 & 0 \\ 0 & \lambda_2 \end{pmatrix}.$$

Change the variables (x, y) and (u, v) to (ξ, η) and (ξ', η') by means of $\binom{x}{y} = T\binom{\xi}{\eta}$, $\binom{u}{v} = T\binom{\xi'}{\eta'}$. Then τ becomes

$$(15.28) \qquad \begin{pmatrix} \xi' \\ \eta' \end{pmatrix} = T^{-1}QT \begin{pmatrix} \xi \\ \eta \end{pmatrix}$$

and T appears in the *canonical form*

$$(15.29) \qquad \begin{cases} \xi' = \lambda_1 \xi \\ \eta' = \lambda_2 \eta. \end{cases}$$

If one sets $z' = \xi + i\eta$, $w' = \xi' + i\eta'$, then the transformation becomes

$$w' = Cz' + D\overline{z'}, \quad C = \tfrac{1}{2}(\lambda_1 + \lambda_2)$$
$$(15.30) \qquad\qquad\qquad\qquad\qquad\qquad C, D, \text{real.}$$
$$D = \tfrac{1}{2}(\lambda_1 - \lambda_2) \neq 0$$

Equation (15.27) becomes $\omega^2 = \overline{D}/D = 1$ so that $\omega = \pm 1$ and the invariant lines are the coordinate axes.

For a nonlinear transformation τ one has

$$(15.31) \quad \tau: w = \alpha z + \beta \bar{z} + \gamma z^2 + \delta z \bar{z} + \epsilon \bar{z}^2 + \cdots$$

which, by a change of variable, can, under the circumstances indicated, be written in the form

$$(15.32) \quad \tau: w' = Cz' + D\bar{z}' + \cdots \qquad C, D \text{ real.}$$

The existence of curves invariant under τ has been studied extensively. One has, e.g.,

Let $|\lambda_1| \neq 1$ *and* $\lambda_2 \neq \lambda_1^p$ *for any positive integer p. Then there exists an analytic arc passing through the origin, tangent to the real z' axis and invariant under τ. If* $|\lambda_2| \neq 1$ *and* $\lambda_1 \neq \lambda_2^q$ *for any positive integer q, then there exists an invariant analytic arc passing through the origin and tangent to the imaginary z' axis. If* $0 < \lambda_2 < 1 < \lambda_1$, *the two invariant arcs are unique.*

In the last case, if $\eta = f(\xi)$ *is the unique invariant arc through the origin tangent to the ξ-axis, and if* $\eta = f_0(\xi)$ *is any function with* $f_0(0) = 0$ *and satisfying a Lipshitz condition in the neighborhood of* $\xi = 0$, *then* $\eta = f(\xi) = \lim_{n\to\infty} f_n(\xi), f_{n+1}(\xi) = \tau f_n(\xi)$.

This notation means that τ transforms the arc $\eta = f_n(\xi)$ into the arc $\eta = f_{n+1}(\xi)$.

15.4. Iteration and orbits. An iterative scheme in two (real) variables is often specified in the form

$$(15.33) \quad \tau: \begin{cases} x_{n+1} = f(x_n, y_n) \\ y_{n+1} = g(x_n, y_n) \end{cases} \qquad n = 0, 1, \cdots.$$

We shall suppose that f and g are regular analytic functions of two variables in an appropriate domain and assume that the domain of definition is such that iteration can be carried out. We can write this in the form

$$(15.34) \quad \begin{cases} x_{n+1} = \tfrac{1}{2}(z_{n+1} + \bar{z}_{n+1}) \\ \qquad = f\left(\dfrac{z_n + \bar{z}_n}{2}, \dfrac{z_n - \bar{z}_n}{2i}\right) \equiv \phi(z_n, \bar{z}_n) \\ \\ y_{n+1} = \dfrac{1}{2i}(z_{n+1} - \bar{z}_{n+1}) \\ \qquad = g\left(\dfrac{z_n + \bar{z}_n}{2}, \dfrac{z_n - \bar{z}_n}{2i}\right) \equiv \psi(z_n, \bar{z}_n). \end{cases}$$

Hence

$$(15.35) \quad z_{n+1} = F(z_n, \bar{z}_n) = (\phi + i\psi)(z_n, \bar{z}_n)$$

is an equivalent form of (15.33) in conjugate variables.

Let us assume that all the iterates $z_0, z_1, \cdots$ under (15.35) lie on an analytic arc C and that the set of points $\{z_n\}$ has a limit point z^* interior to the arc C. There can be at most one such analytic arc. For suppose there were two: call their Schwarz Functions $S(z)$ and $T(z)$. We must have $\bar{z}_n = S(z_n)$ and $\bar{z}_n = T(z_n)$. Also $\bar{z}^* = S(z^*) = T(z^*)$, so that S and T and hence $S - T$ are regular analytic in some neighborhood of z^*. Since $S(z_n) - T(z_n) = 0$ on a set of points with a limit point interior to its region of regularity, the uniqueness theorem for analytic functions tells us that $S(z) - T(z) \equiv 0$.

We shall call C (determined by F and z_0) *the orbit of* $\{z_n\}$.

Let C have the Schwarz Function $T(z)$ and suppose that $\{z_n\}$ has a finite accumulation point on C. Then, $\bar{z}_n =$

$T(z_n)$. Now,

$$z_{n+1} = \bar{T}(\bar{z}_{n+1}) = \bar{T}(\bar{F}(\bar{z}_n, z_n)) = \bar{T}(\bar{F}(T(z_n), z_n)).$$

But, $z_{n+1} = F(z_n, \bar{z}_n) = F(z_n, T(z_n))$ and hence,

$$F(z_n, T(z_n)) = \bar{T}(\bar{F}(T(z_n), z_n)), \qquad n = 0, 1, \cdots.$$

Since the functions $F(z, T(z))$ and $\bar{T}(\bar{F}(T(z), z))$ are regular and the $\{z_n\}$ has a finite accumulation point inside a region of regularity, it follows that we must have

$$(15.36) \qquad F(z, T(z)) \equiv \bar{T}(\bar{F}(T(z), z))$$

identically, or

$$(15.37) \qquad T(F(z, T(z))) = \bar{F}(T(z), z).$$

This functional equation must be satisfied by the Schwarz Function T of the orbit of $\{z_n\}$.

Note also that (15.37) is also a functional equation satisfied by the Schwarz Function $T(z)$ of a curve invariant under the transformation $w = F(z, \bar{z})$ (see (15.25)).

Special cases of this functional equation are noteworthy.

(A) $F(z, \bar{z}) = f(z)$. Then, $\bar{F} = \bar{f}(z)$. This leads to

$$(15.38) \qquad\qquad Tf = \bar{f}T.$$

We have already met up with this equation in (8.10).

Examples. (a) $f(z) = \frac{1}{2}z$, $z_0 = 1$, $\cdots$, $z_n = 2^{-n}$. The orbit is the x-axis. The points $\{z_n\}$ have an interior limit point at $z = 0$. Note that $y = 0$ is not the only analytic arc that passes through the z_i; e.g., $y = \sin(\pi/x)$. However, this arc cannot be continued analytically beyond $x = 0$ so that the limit point would not be interior.

(b) $f(z) = \alpha z$, $\alpha = e^{\pi i\theta}$, $z_0 = 1$. If θ is rational, then there are only a finite number of points among the z_i and no orbit can be determined. If θ is irrational, the points z_n lie everywhere dense on $|z| = 1$. This is the orbit. The functional equation $Tf = \bar{f}T$ becomes the identity $1/\alpha z = \bar{\alpha}/z$.

(c) Let $f(z) = az^p$ where p is a positive integer and $az_0^{p-1} \neq 0, 1$. Then,

$$T(z) = \frac{\bar{z}_0}{z_0^{\omega}} z^{\omega}, \quad \text{where} \quad \omega = \frac{\log (\bar{a}\bar{z}_0^{p-1})}{\log (az_0^{p-1})}$$

satisfies (15.38) and $\bar{z}_0 = T(z_0)$. By (9.7), T is the Schwarz Function of a Bernoulli spiral. All the iterates of z_0 under f lie on it. These spirals are curvewise invariant under f.

If $|a| < 1$ then f maps $|z| \leq 1$ onto $|z| \leq a < 1$. Note that $|f'(z)| = p|a|$, so that for p sufficiently large, we are not dealing with a contraction map. Nonetheless, by the Henrici fixed point theorem, simple iteration of f converges to 0, the unique fixed point.

(B) $F(z, \bar{z}) = g(\bar{z})$. This leads to $\bar{F} = \bar{g}(\bar{z})$; $Tg = \bar{g}T$, or $TgT = \bar{g}$.

(C) Let S be an analytic arc with Schwarz Function $S(z)$ and consider the iteration scheme (15.7) leading to $F(z, \bar{z}) = \frac{1}{2}(z + \bar{S}(\bar{z}))$. Then the Schwarz Function T of the orbit of the iterates satisfies the functional equation

(15.39) $\frac{1}{2}(S(z) + T(z)) = T(\frac{1}{2}(z + \bar{S}(T(z))))$.

Here are some simple consequences of (15.39). If T is the y axis, then $T = -z$, so that $\frac{1}{2}(S - z) = -\frac{1}{2}(z + \bar{S}(-z))$. Therefore $\bar{S}(z) = -S(-z)$. By (8.17″) S is symmetric with respect to the y axis, as is to be expected.

Suppose, following (7.23), we write $S(z) = z - ikz^2 + (-k^2 - (i/3)k')z^3 + \cdots$. Assume $T = az + \cdots$, and comparing the coefficient of z when these are inserted in (15.39), we get $a^2 = 1$. Hence $a = 1$ or $a = -1$. By comparing further coefficients we see that each relation leads to a unique formal solution for T. The selection $a = 1$ leads to $T = S$ which is uninteresting for us. The section $a = -1$ leads to

$$T(z) = -z + 0.z^2 + \frac{i}{3}k'z^3 + \cdots$$

so that the arc T is perpendicular to S and has 0 curvature at the origin.

Suppose we have an arc S and desire to find the orbit of iteration T, at the point z_0 on S. The equation (15.39) suggests the iteration

$$(15.40)\quad T_{n+1}(z) = \tfrac{1}{2}(T_n + (-S + 2T_n(\tfrac{1}{2}(z + \bar{S}T_n(z))))),$$

beginning from the 0th approximation $T_0(z) = \alpha z$ with $\alpha = -S'(0)$. Experience shows this is rapidly convergent.

Example. As S take the arc of the cubic curve

$$\begin{cases} x = t(t - 2) \\ y = t(t - 1)(t - 2), \end{cases}$$

discussed after (8.36). After 8 iterations one obtains $T(z) = -iz + 0z^2 + (-.1875i)z^3 + (.0703125$

$$+ .0703125i)z^4 + (.0546875 - .052734375i)z^5$$

$$+ (-.02758789063 + .1198730469i)z^6$$

$$+ (.02124023437 - .03674316406i)z^7 + \cdots$$

and the 9th iteration coincides with the 8th. The computation was carried out partially by matrix multiplication, as suggested after Equation (8.36).

We turn our attention finally to the special case of the iteration $z_{n+1} = f(z_n)$, f analytic. We suppose that 0 is a fixed point of f: $f(0) = 0$ and that $f'(0) = a \neq 0$. Follow the lead of matrix theory. The high powers of the square matrix A are most easily studied if A can be diagonalized, that is, if there exists an invertible H such that $A = H^{-1}\Lambda H$ where Λ is a diagonal matrix

$$\Lambda = \begin{pmatrix} \lambda_1 & \cdots & 0 \\ & \lambda_2 & \vdots \\ 0 & \cdots & \lambda_q \end{pmatrix}$$

In this case, $A^n = H^{-1}\Lambda^n H$, where

$$\Lambda^n = \begin{pmatrix} \lambda_1^n & \cdots & 0 \\ & \lambda_2^n & \vdots \\ 0 & \cdots & \lambda_q^n \end{pmatrix}$$

By the same token, the function $f(z) = az + \cdots$ will be "diagonalized" if we can find an analytic $H(z) = z + \cdots$, $|z| \leqq \sigma$, such that $f = H^{-1}aH$ or

(15.41) $$Hf = aH,$$

or

(15.42) $$Hf = AH,$$

where A is the "diagonal" function az.

Local iteration is now completely determined insofar as

(15.43) $$z_n = H^{-1}A^nH(z_0)$$

$$= H^{-1}(a^nH(z_0)), \qquad n = 0, 1, \cdots$$

holds for values of z_0 sufficiently close to the origin. If one has $a = \rho e^{i\psi}$ and if one sets

$$(15.44) \quad z_t = z_t(z_0) = H^{-1}(a^t H(z_0)) = H^{-1}(\rho^t e^{it\psi} H(z_0)),$$

then (15.44) serves to define the *fractional* or *continuous iterates* of f. Proper regard must be given to the branches of a^t. These iterates satisfy the functional equation

$$(15.45) \qquad z_u(z_t(z_0)) = z_{u+t}(z_0).$$

If we set $z_t = F(z, t)$ in (15.44), then (15.45) becomes the functional equation

$$(15.46) \qquad F(F(z, u), t) = F(z, u + t)$$

in the three variables z, u, t. This is called the functional equation for *iteration groups*. Under certain conditions it has been established that the general solution can be written in the form $F(z, u) = \zeta^{-1}(a^u \zeta(z))$ or in the equivalent form $F(z, u) = \zeta^{-1}(\zeta(z) + u)$ for an arbitrary invertible function ζ.

15.5. The Schroeder function. Insofar as a complete discussion of matrix diagonalization leads to the Jordan normal form, etc., it is to be expected that a complete discussion of (15.41) is also attended by considerable difficulty. This is indeed the case, as we shall now see, though the difficulties are of a different nature.

The functional equation $Hf = aH$, $a = f'(0)$, is known as the *Schroeder-Koenigs equation*. For short, we shall call H a *Schroeder Function for f*. If we write $\Omega = H^{-1}$, this equation can be written in the alternate form $f\Omega(z) = \Omega(az)$.

Examples. Elementary examples of the Schroeder function can be found by writing $f = H^{-1}aH$ and working backwards from closed-form functions $H(z) = z + \cdots$ whose inverse also has a closed form.

$f(z)$	$H(z)$
$\dfrac{1}{k}((1 + kz)^a - 1)$	$\dfrac{1}{k}\log(1 + kz)$
$\dfrac{1}{k}\log(ae^{kz} - a + 1)$	$\dfrac{1}{k}(e^{kz} - 1)$
$\dfrac{1}{2k}(-1 + (4ak^2z^2 + 4akz + 1)^{1/2})$	$z + kz^2$
$\sin(a \arcsin z)^*$	$\arcsin z$
$\dfrac{az + b}{cz + d}$	See next example.

Example.   Let $w = f(z) = (az + b)/(cz + d)$, $ad - bc \neq 0$, and assume that f has two distinct fixed points $\alpha = (1/2c)((a - d) + \sqrt{(a - d)^2 + 4bc})$ and $\beta = (1/2c)((a - d) - \sqrt{(a - d)^2 + 4bc})$. We have $f'(\alpha) = (ad - bc)/((c\alpha + d)^2)$ and also $(c\alpha + d) = (a - c\beta)$. Hence

$$\begin{pmatrix} 1 & -\alpha \\ 1 & -\beta \end{pmatrix}\begin{pmatrix} a & b \\ c & d \end{pmatrix} = \lambda \begin{pmatrix} f'(\alpha) & 0 \\ 0 & 1 \end{pmatrix}\begin{pmatrix} 1 & -\alpha \\ 1 & -\beta \end{pmatrix} \text{ with } \lambda = a - c\beta.$$

Therefore $H(z) = (z - \alpha)/(z - \beta)$ satisfies $Hf = f'(\alpha) \cdot H$.

---

*If a is a nonnegative integer, this takes the form $(1 - T_a^2((1 - z^2)^{1/2}))^{1/2}$ where T_a is the ath Tschebyscheff polynomial.

The function $(\alpha - \beta)H(z) = (z - \alpha) + \cdots$ is therefore the Schroeder Function for $f(z)$ at the fixed point $z = \alpha$.

The Schroeder identity is often written in the form

$$\frac{w - \alpha}{w - \beta} = K \frac{z - \alpha}{z - \beta}$$

where

$$K = \frac{a + d - \sqrt{s}}{a + d + \sqrt{s}}, \qquad s = (a - d)^2 + 4bc.$$

This exhibits the bilinear transformation in terms of its fixed points. In the (parabolic) case where the fixed points are coincident $(\alpha = \beta = (a - d)/2c)$, the bilinear transformation can be written as

$$\frac{1}{w - \alpha} = \frac{1}{z - \alpha} + \frac{2c}{a + d}, \qquad \text{if} \quad a + d \neq 0.$$

The theory of the Schroeder Function is at the present writing (1973) incomplete. We shall give a brief summary of what is known.

If n is a positive or negative integer, the Schroeder Function for f is also the Schroeder Function for f^n (the nth iterate of f). For if $f(z) = sz + \cdots$, then $f^n(z) = s^n z + \cdots$. Now $Hf^2 = Hff = sHf = ssH = s^2H$ so that H is the Schroeder Function for f^2, etc. For more general values of n, see the remark after (15.44).

If H is the Schroeder Function for f and if $h(z) = z + \cdots$, then Hh^{-1} is the Schroeder Function for the conjugate $g = hfh^{-1}$. For $(Hh^{-1})g = Hh^{-1}(hfh^{-1}) = (Hf)h^{-1} = sHh^{-1}$.

If H is the Schroeder Function for f then $\bar{H}$ is the Schroeder Function for $\bar{f}$.

THEOREM. *Let* $f(z) = az + bz^2 + cz^3 + \cdots$ *be a formal power series and suppose that* $a \neq 0$ *and* a *is not a root of unity* $(a^p \neq 1)$. *Then there is a unique formal power series* $H(z) = z + \cdots$ *which satisfies the Schroeder equation* $Hf = aH$.

Proof: Set $H(z) = z + \alpha z^2 + \beta z^3 + \cdots$ and insert f in H. Compare the coefficients of Hf against those of aH. One may solve for $\alpha, \beta, \cdots$ successively since the coefficients of the leading term in $\alpha, \beta, \cdots$ will be $a^q - a \neq 0$.

One requires, of course, more than a formal power series solution.

CASE I. $|a| \neq 1, 0$. In this case, there exists a unique Schroeder Function. The functional equation $Hf = aH$ suggests the iteration $H_{n+1} = (1/a)H_n f$ and this iteration with $H_0 \equiv z$ converges.

CASE II. $|a| = 1$. This is the most difficult and in some ways the most interesting case. This case has been called by some authors "the function-theoretic center problem" because of certain relations to stability problems.

CASE IIa. a is a pth root of unity: $a^p = 1$, for some $p = 1, 2, \cdots$. In this case there is a Schroeder Function for $f(z)$ if and only if $f^p = ff \cdots f \equiv z$. There is no uniqueness of the solution.

CASE IIb. a is not a pth root of unity (for any $p = 1, 2, \cdots$). In this case, formal substitution of the series $H(z) = z + \cdots$ in (15.41) will lead to a succession of equations which can be solved uniquely. Thus, there is always a formal solution. This formal solution may converge to a proper solution or it may diverge. The set of values of a for which there is a convergent $f(z) = az + \cdots$ whose Schroeder series diverges is dense on $|a| = 1$.

CASE IIb$_1$. There exists a constant $k > 0$ such that

$$(15.47) \qquad \big| \log | a^n - 1 | \big| \leq k \log n, \qquad n = 2, 3, \cdots.$$

In this case, the Schroeder series converges to a proper Schroeder Function. Condition (15.47) is fulfilled almost everywhere on $|a| = 1$, but given an a, it is extremely difficult to tell whether it fulfills (15.47).

We now give the proofs of several of the preceding statements.

THEOREM. *Let $f(z)$ be analytic in a neighborhood of $z = 0$, $f(0) = 0$, and suppose that $f'(0) = s$ with $|s| \neq 0, 1$. Then there exists a unique Schroeder Function for $f(z)$. It can be found from the iteration $H_{n+1} = (1/s) H_n f$ beginning with $H_0(z) \equiv z$.*

Proof: Case 1. $0 < |s| < 1$.

Starting from a z_0, define the sequence $\{z_n\}$ by $z_{n+1} = f(z_n)$, $n = 0, 1, \cdots$. As in the previous theorem, if $|z_0| \neq 0$ is sufficiently small, $z_n \neq 0$ and $\lim_{n \to \infty} z_n = 0$. Since

$$\frac{z_{n+1}}{z_n} = \frac{f(z_n)}{z_n} = s + b z_n + c z_n^2 + \cdots,$$

it follows that $\lim_{n \to \infty} (z_{n+1}/z_n) = s$. Write f^n for the nth iterate of f, and consider the function $H_n(z_0) = (f^n(z_0)/s^n)$. This is an analytic function of z_0 in a neighborhood of the origin. Now

$$H_n(z_0) = \left(\frac{z_n}{s z_{n-1}} \right) \left(\frac{z_{n-1}}{s z_{n-2}} \right) \cdots \left(\frac{z_1}{s z_0} \right) z_0, \qquad (z_0 \neq 0),$$

$H_n(0) = 0$. The general term of this product is z_{n+1}/sz_n. In the expansion $f(z)/z = s + bz + cz^2 + \cdots$, let the term with the first non-zero coefficient be az^h, $h \geqq 1$, $a \neq 0$. Then, we can write $z_{n+1}/sz_n = f(z_n)/sz_n = 1 + w_n$ where $w_n = az_n^h/s + \cdots$. Consider now

$$\frac{w_{n+1}}{w_n} = \left(\frac{az_{n+1}^h}{s} + \cdots\right) \div \left(\frac{az_n^h}{s} + \cdots\right)$$

so that $\lim_{n\to\infty} (w_{n+1}/w_n) = \lim_{n\to\infty} (z_{n+1}/z_n)^h = s^h$. Since $|s| < 1$, it follows that $\sum_{n=1}^{\infty} |w_{n+1}/w_n| < \infty$, uniformly for z_0 in closed subsets of $0 < |z| < \rho$.

By general theorems on infinite products, it follows that $H_n(z)$ converges uniformly to an analytic function $H(z)$. The difficulty at $z = 0$ is only apparent. We have $sH_{n+1}(f(z)) = H_n(z)$, so that in the limit $Hf = sH$. Finally, $H_n(z) = f^n(z)/s^n = z + \cdots$, so that $H'(0) = 1$. H is therefore the Schroeder Function for f.

Case 2. $|s| > 1$. If $w = f(z) = sz + \cdots$, consider the inverse function $f^{-1}(z) = (1/s)z + \cdots$. Since $0 < |1/s| < 1$, f^{-1} has a Schroeder Function $H: Hf^{-1} = (1/s)H$. Hence $H = (1/s)Hf$ or $Hf = sH$, so that H is also the Schroeder Function for f.

THEOREM. *Let p be an integer $\geqq 1$ and let $a^p = 1$. Then $f(z) = az + \cdots$ has a Schroeder function $H(z) = z + \cdots$ if and only if $f^p \equiv z$. In this case, there is no uniqueness and a general solution can be given which depends upon an arbitrary analytic function.*

Proof: Suppose that H is a Schroeder function for f. Then $Hf = aH$ so that $(HfH^{-1})(z) \equiv az$. The pth power in the sense of functional composition is $Hf^pH^{-1} = a^pz \equiv z$. Hence $f^p \equiv z$.

Conversely, let $f^p \equiv z$. Let $g(z)$ be an arbitrary analytic function defined in a neighborhood of $z = 0$. Form

$$H(z) = \sum_{k=0}^{p-1} a^{-k} g f^k.$$

Now

$$Hf = \sum_{k=0}^{p-1} a^{-k} g f^{k+1} = a \sum_{r=1}^{p} a^{-r} g f^r = a \sum_{r=1}^{p-1} a^{-r} g f^r + ag = aH.$$

Thus H is a Schroeder function for f.

Note also that any Schroeder function can be put into this form by selecting $g = (1/p)H$. For since $Hf = aH$, $Hf^k = a^k H$. Therefore

$$\sum_{k=0}^{p-1} a^{-k} \frac{1}{p} Hf^k = \sum_{k=0}^{p-1} a^{-k} \frac{1}{p} a^k H = H.$$

If the equation $Hf = AH$ is given the matrix interpretation through Equation (8.31), the matrix A is the diagonal

$$A = \begin{pmatrix} a & & & & \\ & a^2 & & & \\ & & a^3 & & \\ & & & \cdot & \\ & & & & \cdot \\ & & & & & \cdot \end{pmatrix}.$$

Hence, the numbers $a, a^2, \cdots$ may be thought of as the "eigenvalues" of $f(z)$. Case I and IIb are cases where the "eigenvalues" are all distinct and there is at least a formal solution to the problem of diagonalization.

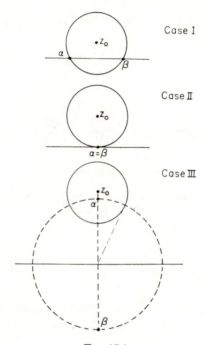

FIG. 15.1

Case IIb_1, due to C. L. Siegel, lies very deep and is allied to the "method of small divisors."

Example. Let $z_0 = x_0 + iy_0$. The circle $|z - z_0| = r$ has the Schwarz Function $S(z) = r^2/(z - z_0) + \bar{z}_0$. The fixed points α, β of this bilinear transformation are

$$x_0 \pm \sqrt{r^2 - y_0^2} \text{ if } y_0^2 \leqq r^2$$

or

$$x_0 \pm i\sqrt{y_0^2 - r^2} \text{ if } y_0^2 > r^2.$$

We can therefore distinguish three cases (see Fig. 15.1).

I. The circle intersects the x-axis in two distinct real points; they are the fixed points α, β. II. The circle is tangent to the x-axis at α: the two fixed points coincide $\alpha = \beta$. III. The circle does not intersect the axis. The fixed points are inverse in the circle and are obtained as in the accompanying construction. (See Fig. 15.1,III.)

In cases I and III, $S(z)$ has a Schroeder Function, as in the example on page 194. In Case II, $\alpha = \beta = x_0$, $S'(x_0) = 1$, a first root of unity, so that $S(z)$ has no Schroeder Function by the previous theorem.

Example (Numerical). Find the Schroeder Function for $f(z) = \frac{1}{2}z + z^2$. We used the solution $H(z) = \lim_{n \to \infty} 2^n f^n(z)$ and the matrix representation (8.5.10). After 39 iterations we obtained

$$H(z) \approx z + 4z^2 + 10.66666667z^3 + 27.42857143z^4$$
$$+ 63.39047619z^5 + 147.8586789z^6$$
$$+ 328.7745836z^7 + 726.6408688z^8 + \cdots.$$

The 40th iteration yielded the same values to 10 significant figures.

Relevant to the case $|a| = 1$ is the following characterization of stability.

THEOREM. *The mapping* τ: $z' = f(z) = az + bz^2 + \cdots$ *is stable at the fixed point* $z = 0$ *if and only if* $|a| = 1$ *and f has a Schroeder function* $H(z) = z + \cdots$.

Proof—Sufficiency. Assume that $|a| = 1$ and H exists with $Hf = aH$. For all sufficiently small r, the image of the disc D: $|w| < r$ under $z = H^{-1}(w)$ is a circle-like region B containing $z = 0$. If $z \in B$, then $f(z) = fH^{-1}(w) = H^{-1}(aw)$. Now, if $w \in D$, then $aw \in D$. Hence, $H^{-1}(aw) =$

$f(z) \in B$. Similarly, $f^{-1}(z) = f^{-1}H^{-1}(w) = (Hf)^{-1}(w) = (aH)^{-1}(w) = H^{-1}(w/a)$. Now $w/a \in D$, so that $f^{-1}(z) \in B$. Thus, $\tau(B) = B$ and B is an invariant neighborhood of $z = 0$. Any neighborhood of $z = 0$ contains such a B and hence τ is stable.

Necessity. If ρ is the radius of convergence of $az + bz^2 + \cdots$, by the definition of stability, the disc $|z| < \rho$ contains an invariant neighborhood B of the fixed point $z = 0$ $(\tau(B) = B)$. We shall show that we can find an invariant neighborhood which is also simply connected. Let C be an open disc contained in B and containing $z = 0$. For each $n = 0, \pm1, \pm2, \cdots, \tau^n C \subset B$ and contains $z = 0$. Hence, the set $D = \cup_{n=-\infty}^{\infty} \tau^n C$ is a connected invariant neighborhood of $z = 0$. If D is not simply connected adjoin to D all the points that lie interior to simple closed curves Γ lying in D. Call the enlarged set D^*. Then D^* is clearly simply connected. Furthermore, it is invariant under τ. For, $\tau D = D$. Now take $z \in D^* - D$. It lies interior to some curve Γ. Hence, $\tau(z)$ is interior to $\tau(\Gamma)$. But $\tau(\Gamma) \subset B$ so its interior is contained in D^*. Therefore $\tau(z) \in D^*$. Similarly for $\tau^{-1}(z)$.

By the Riemann mapping theorem, D^* can be mapped by $w = M(z) = z + \cdots, z = M^{-1}(w) = w + \cdots$ one-to-one conformally onto a disc $\Delta: |w| < \sigma$. Consider the composite function $T(w) = MfM^{-1}(w)$. For $w \in \Delta$, $T^{\pm1}(w) \in D$. Hence, T maps Δ onto itself conformally with $w = 0$ as a fixed point. By a well-known theorem of conformal mapping (see, e.g., Ahlfors [A1] p. 136) it follows that there is an α, with $|\alpha| = 1$, such that $T(w) = \alpha w$ (i.e., T is a rotation). Thus, $T(w) = MfM^{-1}(w) = \alpha w$, $|\alpha| = 1$. This implies that $a = \alpha$ and $|a| = 1$. Also $Mf(z) = aM(z)$, so that $M(z)$ is a Schroeder Function for f.

We note in passing from the first part of the proof that the images of discs $|w| < r$ under H^{-1} are invariant neighborhoods B of $z = 0$ under f. If $S(z)$ designates the Schwarz Function of the boundary of B, then from (8.4), $S(z) = \bar{H}^{-1}(r^2/H(z))$ so that we have the identity $\bar{H}S(z) \cdot H(z) = r^2$.

Suppose now that $f(z) = f'(0)z + \cdots$ is a mapping in which, for simplicity, we assume that $f'(0)$ is real. Assume that f has a unique Schroeder Function $H(z) = z + \cdots$ with $Hf = f'(0)H$. Suppose that the analytic arc S with Schwarz Function $S(z) = \alpha z + \cdots, |\alpha| = 1$ is curvewise invariant under f (or is an orbit for $\{z_n\}$ with $z_0 \in S$). What is the relationship between S and H? From the above relationship, $\bar{H}\bar{f} = f'(0)\bar{H}$, and hence $\bar{H}\bar{f}S = f'(0)\bar{H}S$. Since S is invariant under f, $\bar{f}S = Sf$ so that $\bar{H}Sf = f'(0)\bar{H}S$. Now $\bar{H}S = \alpha z + \cdots$, so that $(1/\alpha)\bar{H}S = z + \cdots$. Therefore $\{(1/\alpha)\bar{H}S\}f = f'(0)\{(1/\alpha)\bar{H}S\}$, and $(1/\alpha)\bar{H}S$ is also the Schroeder Function for f. This means that $(1/\alpha)\bar{H}S = H$ so that

$$(15.48) \qquad\qquad S = \bar{H}^{-1}\alpha H.$$

Conversely, for any α with $|\alpha| = 1$, and any invertible $H(z) = az + \cdots, a \neq 0, \bar{H}^{-1}\alpha H$ is the Schwarz Function of an arc. For if we take

$$f(z) = H^{-1}(z/\sqrt{\alpha}), \qquad \bar{f} = \bar{H}^{-1}(z/\sqrt{\bar{\alpha}})$$

$f^{-1}(z) = \sqrt{\alpha}H(z)$ so that

$$\bar{f}f^{-1}(z) = \bar{H}^{-1}\frac{\sqrt{\alpha}H(z)}{\sqrt{\bar{\alpha}}} = \bar{H}^{-1}\alpha H(z).$$

THEOREM. *Suppose that $f(z) = az + \cdots$ has a Schroeder Function $H(z) = z + \cdots$, $Hf = aH$. For points $z_0 \neq 0$*

sufficiently near the origin, the continuous real iterations $z_t(z_0)$ *of* z_0 *under* f *trace out an arc which is invariant under* f.

Proof: From (15.44), the continuous iterates of z_0 are given by $H^{-1}(a^t H(z_0))$, t real. Write this as $H^{-1}Q$ with $Q(t) = a^t H(z_0)$. From (8.1), this curve has Schwarz Function $S = \overline{H^{-1}Q}(H^{-1}Q)^{-1} = \bar{H}^{-1}\bar{Q}Q^{-1}H$. Now

$$\bar{Q}Q^{-1}H = \overline{H(z_0)}\,(H(z)/H(z_0))^{(\log \bar{a}/\log a)}$$

and

$$\bar{Q}Q^{-1}aH = \overline{H(z_0)}\,(aH(z)/H(z_0))^{(\log \bar{a}/\log a)}.$$

Now,

$$a^{(\log a/\log a)} = e^{(\log a/\log a)\log a} = e^{\log \bar{a}} = \bar{a}$$

so that $\bar{Q}Q^{-1}aH = \bar{a}\bar{Q}Q^{-1}H$. Therefore

$$\bar{Q}Q^{-1}Hf = \bar{H}\bar{f}\bar{H}^{-1}\bar{Q}Q^{-1}H$$

or

$$\bar{H}^{-1}\bar{Q}Q^{-1}Hf = \bar{f}\bar{H}^{-1}\bar{Q}Q^{-1}H,$$

$$Sf = \bar{f}S.$$

Therefore, by (8.10), S is invariant under f.

A (partial) solution to the bisection problem can now be given in terms of the Schroeder Function. Given the arc S passing through $z = 0$, to bisect the curvilinear angle formed by S and the x-axis, we must solve the functional equation $TT = S$.

THEOREM. *Let the Schwarz Function* $S(z) = \alpha z + \cdots$, $|\alpha| = 1$, *possess a Schroeder Function* $H(z) = z + \cdots$, $HS = \alpha H$. *Suppose that the function* $T = H^{-1}(\pm\sqrt{\alpha})H$ *is the Schwarz Function of an arc* T. *Then* T *is a bisector. Conversely, suppose that* α *is not a root of unity and the*

curvilinear angle has a bisector T. Then T is given by
$T = H^{-1}(\pm\sqrt{\alpha})H$.

Proof—First part. $TT = H^{-1}(\pm\sqrt{\alpha})HH^{-1}(\pm\sqrt{\alpha})H$

$$= H^{-1}\alpha H = H^{-1}HS = S.$$

Second part. If α is not a root of unity, then, as we have seen in Chapter 8, the functional equation $TT = S$ has precisely two formal solutions. But the power series $H^{-1}(\pm\sqrt{\alpha})H$ provide two convergent solutions so that one of them must be T.

Example. Let S be the arc of a circle, passing through the origin and intersecting the x-axis at $x = \beta \neq 0$. Let the center of the circle be z_0 so that $\beta = 2\,\mathrm{Re}\,z_0 = z_0 + \bar{z}_0$ and the Schwarz Function for S is $S(z) = \bar{z}_0 z/(z - z_0)$. The fixed points of $S(z)$ are $z = 0$, $z = \beta$, so that $H(z) = \beta z/(\beta - z) = z + \cdots$ is the Schroeder Function for $S(z)$. We have $S'(0) = -\bar{z}_0/z_0$ so that if we set $z_0 = \rho e^{i\theta}$, then $\theta \neq \pm\pi/2$. $S'(0) = -e^{-2i\theta}$. Then $\sqrt{S'(0)} = e^{-i(\theta\pm\pi/2)} = a \neq 1$. We compute:

$$T = H^{-1}aH = \frac{a\beta z}{\beta + (a-1)z}.$$

T is easily seen to be the Schwarz Function of the circles passing through the origin and center at $z_1 = \beta/(1-a)$. These are the bisectors.

If $z_0 = x_0 + iy_0$ and if $x_0 \to 0$ with $y_0 \neq 0$ fixed, then $\beta \to 0$ and $z_1 \to 2iy_0$. We have already seen this in Chapter 8. The existence of a Schroeder Function for S is therefore not necessary for a bisector to exist.

CHAPTER 16

DICTIONARY OF FUNCTIONAL RELATIONSHIPS

Interpretation	Functional Equation	Equation No.
S is involutory	$S\bar{S} = \bar{S}S = I$	(6.20)
Parametric representation	$S = \bar{f}f^{-1}$	(8.1)
Representation by mapping function	$S = \bar{m}(1/M)$	(8.4)
Image of arc under analytic map	$S_C f = \bar{f}S_B$	(8.7)
Analytic continuation	$f = \bar{S}_C \bar{f}S_B$	(8.7')
Curvewise invariance of S under f	$Sf = \bar{f}S$	(8.10)
Reflection of C in x-axis	$S_{\bar{c}} = \bar{S}_c$	(8.12)
Symmetry in real axis	$\bar{S} = S$	(8.13)
U is reflection of T in S	$U = S\bar{T}S$	(8.15)
Symmetry of two arcs	$S\bar{T} = T\bar{S}$	(8.16')
Schroeder Function	$Hf = aH$	(15.41)
Orbit	$\bar{H}T = aH, \, |a| = 1$	(15.48)

CHAPTER 17

BIBLIOGRAPHICAL AND SUPPLEMENTARY NOTES

Chapter 1: The notes of L. B. Rall were subsequently published as Rall [**R1**].

Chapters 2, 3, 4: For plane geometry worked through complex numbers see, for example, W. B. Carver [**C2**], Morley and Morley [**M5**], H. B. Eves [**E1**], I. M. Yaglom [**Y1**]. H. Schwerdtfeger [**S5**] presents inversive geometry with emphasis on matrix methods. There are many beautiful things in these books which are accessible to elementary courses.

Carver [**C2**] uses a somewhat different definition of the clinant.

(2.1): Admittedly, by restricting x and y to real values, certain unifying principles are lost. The author has used this restriction to control the scope of the essay. Incidentally, *three* complex planes corresponding to three complex number systems with commutative and associative multiplication have been studied. They are (1) $x + iy$, $i^2 = -1$; the normal-complex plane, (2) $x + \epsilon y$, $\epsilon^2 = 0$; the dual-complex plane, (3) $x + jy, j^2 = 1$, the abnormal-complex plane. For applications of dual-complex numbers to the geometry of oriented circles, see Yaglom [**Y1**].

Apropos the "beautiful irrelevance" of the nine-point

circle theorem, it is interesting to note that this theorem has often been made the scape-goat for the imagined sins of the mathematical education of past generations. Thus, at a recent conference (see, "Panel Discussions", Second National Mathematics Conference, Arya-Mehr University of Technology, Teheran, Iran, March, 1971, pp. 17–30), Prof. Jean Dieudonné cites the nine-point circle as an example of a "silly theorem" about a triangle which once proved is never needed any more. He goes on to suggest that the simplicity of the three-dimensional orthogonal group would, for example, be far more relevant to elementary education. By way of rebuttal, Prof. John McCarthy, a computer scientist, observed that neither of these theorems ever motivated him very much.

The abstract problem of what mathematics each generation should transmit therefore remains unanswered. Mathematics is open-ended, and attempts to discover its essential and eternal core result in failure.

Chapters 5, 6: Reflection in an analytic arc was introduced in 1870 by H. A. Schwarz in his paper *"Über die Integration der partiellen Differentialgleichung $\Delta u = 0$ unter vorgeschriebenen Grenz- und Unstetigkeitsbedingungen."* It is reprinted in Schwarz [S4].

Reflection in straight lines and circles is presented in most books on complex variables. The general case is dealt with infrequently. The geometry of Schwarzian reflection has been studied extensively by E. Kasner and his students. Kasner's complete bibliography can be found in Jesse Douglas' article [D8]. Kasner's papers are terse and reveal, I think, less than the man.

The Schwarz Function $S(z)$, *qua* analytic function, was named and stressed in Davis and Pollak [D6]. Applications were given to problems of complex moments and analytic continuation. A more extensive list of elementary Schwarz Functions is found in Davis [**D5**].

Despite the fact that the term 'Schwarz Function' is used in the theory of automorphic functions in a totally different sense, there is ample justification to allow the term in its present sense.

On the implicit function theorem for analytic functions leading to the representation (5.10′), see, e.g., Hille [**H4**], vol. I, p. 269.

(6.20), (6.23): The $n \times n$ matrices S satisfying the equation $\bar{S}S = S\bar{S} = I$ have been called *circle matrices* by some authors. See Jacobsthal [**J3**] for a study.

For trioperational algebra, see K. Menger [**M2**].

The representation of linear fractional transformations by 2×2 matrices goes back to Cayley.

One extension of inversive geometry to n complex variables is known as *symplectic geometry* and is expounded in Siegel [**S7**]. It is intimately related to the study of Hamiltonian dynamical systems. See Siegel and Moser [**S6**].

Chapter 7. The reader who wishes to pursue the topic of Schwarzian reflection from the point of view of differential geometry (conformal symmetry) should consult the works of E. Kasner, G. Pfeiffer, and J. de Cicco [**K4**], [**P3, 4**]. Formula (7.22), worked out essentially for $\overline{S(z)}$, is given in Kasner [**K3**]. See also Davis [**D5**].

For applications of the Schwarzian derivative to conformal mapping, see, e.g., Hille [**H4**], vol. II, pp. 375–380, Nehari [**N1**].

Chapter 8. Davis and Pollak [**D6**]. Interest in the functional iteration of analytic functions and the topic of permutable functions led to the spelling out of these and related properties for the Schwarz Function. The Schwarz

Function as a source of various functional equations seems to be overlooked by authors on the latter topic. For permutable functions in the real, see Kuczma [**K7**]. For permutable functions and iterations, the work of I. N. Baker [**B1**] is of great importance.

The composite function $f'f^{-1}$ appearing in (8.1′) essentially twice is crucial to numerous discussions in differential geometry. If A a matrix whose elements are functions of the variable u, recent writers have designated $C(A) = (dA/du)A^{-1}$ as the *Cartan matrix* of A. The identity $C(AB) = C(A) + AC(B)A^{-1}$ is said to contain a good deal of differential geometry.

For quasi-conformal mappings, see Lehto and Virtanen [**L1**].

For a theory of reflection in certain classes of closed Jordan curves, see Sloss [**S9**].

For reflection laws for various P.D.E.'s, see Lewy [**L3**], Sloss [**S8, 9**].

For the matrix representation of functional composition and the identity (8.36), see I. Schur [**S3**], E. Jabotinsky [**J2**]. These authors apply the identities to the Faber polynomials in terms of which analytic functions in general regions may be expanded. The proof of the Lagrange-Bürmann Theorem, emphasizing that it is a formal algebraic identity, is due to Henrici [**H2**].

On curvilinear angles and the bisection problem, see E. Kasner [**K1**], [**K3**], Pfeiffer [**P3**], [**P4**], [**P5**]. In [**K2**], Kasner treats the problem of the conformal equivalence of analytic arcs with singularities.

Pfeiffer remarks that G. D. Birkhoff took a lively interest in the two problems of this section. Although he appears not to have published on the topic, he directed one thesis on it (L. T. Wilson, 1915). Prof. Garrett Birkhoff recalls that his father emphasized the distinction between "formal solutions" and "convergent solutions," a distinction which is critical here. See G. D. Birkhoff, *Collected Mathematical Works*, vol. 1, p. 519 ff.

Chapter 9: This chapter formed the basis for an address by the author to the Northeastern Section of the MAA, Fall 1969, Wheaton College, Norton, Massachusetts. The simple form of the Schwarz Function for an equiangular spiral is particularly striking and would undoubtedly have pleased J. Bernoulli, the discoverer of many of the regenerative properties of these curves. The leading question in the title of the chapter can be answered in several different ways. The answer given appears to be the most interesting.

For more on curves invariant under Möbius transformation, see Schwerdtfeger [S5].

For the classification of the singularities of O.D.E.'s see, e.g., Hurewicz [H5].

Chapter 10: See Pólya [P7], Davis and Pollak [D6] for the case of $S(z)$ rational. Pólya's proof of the inequality (10.2) is based upon a theorem of H. Bohr which is of interest in its own right. Pólya makes use of the inequality to prove the following theorem on entire functions of finite order: if g and h are entire and gh is entire of finite order, then either (a) h is a polynomial and g is of finite order or (b) h is of finite order and g is of zero order. Gross [G4] may be profitably consulted.

In the same spirit as the theorem on the rationality of $S(z)$ is the following result which is constantly rediscovered. Let $w = f(z)$ be a schlicht map of $|z| \leqq \infty$ into itself. Suppose that every circle or straight line is mapped into a circle or a straight line. Then $f(z)$ is a Möbius transformation $f(z) = (az + b)/(cz + d)$ or the conjugate of such. See, e.g., Schwerdtfeger [S5], p. 106.

On the real analogues of $S\bar{S} = I$. The equation $f^n(x) = x$, $n > 1$, $f^n = ff \cdots f$, is known as the *Babbage equation*. For a discussion of real solutions, see Kuczma [K7], Chap. 15. The only meromorphic solutions of the Babbage equation are of the form $(ax + b)/(cx + d)$.

Chapter 11. The operators $\partial/\partial z$, $\partial/\partial \bar{z}$ are found many places, e.g., Bergman [B3], Garabedian [G1], Vekua [V1], Keckic [K5], Lehto and Virtanen [L1].

Bochner [B5], Vekua [V1] give quite general sufficient conditions for Green's Theorem. Formula (11.31) was already known to D. Pompeiu. On the singularities of $S(z)$ see Davis and Pollak [D6]. Formula (11.34) was originally derived by I. J. Schoenberg and T. Motzkin and is reported in Schoenberg [S2]. Their derivation uses the Hermite-Genocchi remainder formula. It was derived independently by Grunsky [G5], [G6], [G7] who makes applications to second order ordinary differential equations. See also Davis [D2], [D5]. For (11.40) see Davis [D5]. For (11.48) and related identities, see Davis [D3], [D5]. For (11.38) see Grunsky [G6]. For the Darboux-type expansion in the Example, see Grunsky [G5].

The general solution of $\Delta^n F = 0$ is easily represented in terms of $2n$ analytic functions. See Vekua [V1], Par. 35.

On the analytic continuation of harmonic functions see Garabedian [G2], p. 646. On the Cauchy problem, see Henrici [H1].

For the theory of quasi-conformal mappings, consult, e.g., Lehto and Virtanen [L1].

The two-circle method in photoelasticity is due to Lewis and Pollak [L2]. See this paper for additional identities, some involving the Schwarz Functions of confocal ellipses. See also Neményi and Sáenz [N2] for further applications of complex variable methods to plane stress fields.

Chapter 12. For standard material on elementary fluid mechanics, see, e.g., Milne-Thompson [M3].

Chapter 13. The method of images for regions bounded by a straight line, circle, plane or sphere is commonly discussed in books on partial differential equations or complex variables. The case of a region with an algebraic boundary where the branches of the Schwarz Function have certain group properties (not discussed in this essay) derives from D. A. Grave [G3]. Related material on the solution of the fundamental problems of potential theory via Cauchy integrals and the Plemelj formula can be found, e.g., in L. C. Woods [W3]. This can serve as an introduction to the methods of Muskhelishvili and his school in problems in plane elasticity.

Chapter 14. Davis [D4], [D5]. Walsh's theorem will be found on p. 40 of Walsh [W1]. This theorem is related to one of M. and F. Riesz on measures that are orthogonal to analytic functions on the unit disc. For $L^2(B)$ see, e.g. Bergman [B2], Davis [D1].

On the complete packing of a region B by disks, see O. Wesler [W2], where a theorem of wide generality is proved. The use in complex variable theory of packings of regions by disks goes back at least as far as the work of D. Pompeiu. They are a source of many pathological examples.

The divergence of $\sum r_n$ for complete packings is due to S. Mergelyan (see Wesler [W2]). The convergence of the series $\sum_{n=1}^{\infty} r_n^{\alpha}$ has been investigated by Z. A. Melzak

[**M1**] and others. A considerable literature has grown up here.

Haber's result is this. Let $w(x)$ be a *modulus of continuity*; i.e., $w(x)$ is real-valued, defined on $(0, \infty)$, monotone increasing and with $\lim_{x \to 0^+} w(x) = 0$. Let $F(w)$ be the family of all complex-valued functions f defined on $[a, b]$ and having the property that for some $C = C(f)$,

$$|f(x_1) - f(x_2)| \leqq C(f) \cdot w(|x_1 - x_2|), \quad x_1, x_2 \in [a, b].$$

Then F has a simple quadrature.

Chapter 15. Some of the new material in this section formed the basis of a talk by the author at the Oxford University Computation Center, Oxford, England, January, 1971.

The contraction mapping theorem can be found in most books on functional analysis and most recent books on numerical analysis.

For Henrici's fixed point theorem, see Henrici [**H3**]. Prof. Henrici pointed out to me that the conditions of his theorem may be weakened. Thus, it suffices for f to be analytic in a simply connected region R and the closure of $f(R)$ to be contained in R. The proof requires merely the Riemann mapping theorem itself. Furthermore, if R is multiply connected, one may work with the Green's function for R and obtain a similar result.

Condition (15.3) can be found in Ostrowski [**O1**], Chap. 18. Here is a more general formulation: suppose that $G: R^n \to R^n$ from real n-dimensional space and has x^* as a fixed point: $x^* = Gx^*$. Suppose, further, that G has a Fréchet derivative $G'(x^*)$ at x^* and that the spectral radius $\rho(G'(x^*)) < 1$. Then x^* is attractive for the iteration $x_{n+1} = Gx_n$.

For invariant curves in the real, see Montel [M4], Kuczma [K7]. The existence of invariant curves is of great importance in celestial mechanics. See Siegel and Moser [S6]. Relationships to the Schwarz Function would seem to be desirable. For iteration and Schroeder Functions, see Montel [M4], Kuczma [K7], Siegel and Moser [S6].

The proof of Case IIb1 (Siegel's Theorem) given by Moser in Siegel and Moser [S6] is an excellent illustration of the Kolmogoroff-Arnold-Moser techniques of "quadratic convergence." Kuczma has many references to papers on analytic iterates. For further material on the continuous iteration of Möbius transformations, see Schwerdtfeger [S5], Chap. II, Sec. 10. For iterates of entire functions see Gross [G4], Chaps. 8, 9.

An extensive computer investigation of non-linear iterations is given in Stein and Ulam [S10].

By way of underlining the necessity of computer work, these authors cite the following "simple" example. Consider the one variable iteration

$$y_{n+1} = w_n(3 - 3w_n + \sigma w_n{}^2);$$
$$w_n = 3y_n(1 - y_n), \qquad 0 < y_0 < 1.$$

If $\sigma = .99004$, then for "almost all" y_0, the sequence $\{y_n\}$ converges to a cycle of order 14. If $\sigma = .99005$, it converges to a cycle of order 28. If $\sigma = .99008$, no cyclic behavior was observed within 5×10^5 iterations.

I have seen many things on the computer scope which are exciting, even startling, and which could do much to restore the graphical image as a proper object of mathematical study. The computer scope has remarkable capabilities of animation, and the dynamic figures which emerge are—to steal a line from Descartes—as far removed from old-fashioned static figures as are the orations of Cicero from the simple *ABC*'s. They cannot be exhibited in books, they must be experienced as movies.

I had wanted—particularly in this chapter—to employ the computer scope liberally to illuminate many of the points made here and to suggest new problems. But this is for the future.

18. BIBLIOGRAPHY

[A1] L. H. Ahlfors, *Complex Analysis*, McGraw-Hill, New York, 1966.

[B1] I. N. Baker, Permutable Power Series and Regular Iteration, *J. Australian Math. Soc.*, vol. II, part 3 (1962) 265–294.

[B2] S. Bergman, *The Kernel Function and Conformal Mapping*, New York, 1950.

[B3] ———, *Integral Operators in the Theory of Linear Partial Differential Equations*, Springer, New York, 1969.

[B4] M. Bôcher, *Introduction to Higher Algebra*, Macmillan, New York, 1935.

[B5] S. Bochner, Green-Goursat Theory, *Math. Zeit.*, 63 (1955) 230–242.

[C1] C. Carathéodory, *Conformal Representation*, Cambridge University Press, Cambridge, 1941.

[C2] W. B. Carver, The Conjugate Coordinate System for Plane Euclidean Geometry, No. 5 of the *Slaught Memorial Papers; Amer. Math. Monthly*, No. 9, 63 (1956) November.

[C3] J. L. Coolidge, *Geometry of the Complex Domain*, Oxford University Press, Oxford, 1924.

[D1] P. J. Davis, *Interpolation and Approximation*, Ginn-Blaisdell, Waltham, Mass., 1963.

[D2] ———, Triangle Formulas in the Complex Plane, *Mathematics of Computation*, 18 (1964) 569–577.

[D3] ———, Simple Quadratures in the Complex Plane, *Pacific J. Math.*, 15 (1965) 813–824.

[D4] ———, Additional Simple Quadratures in the Complex Plane, *Aequationes Mathematicae*, 3 (1969) 149–155.

[D5] ———, Double Integrals Expressed as Single Integrals or Interpolatory Functionals, *J. Approximation Theory*, 5 (1972) 276–307.

[D6] P. J. Davis and H. Pollak, On the Analytic Continuation of Mapping Functions, *Trans. Amer. Math. Soc.*, 87 (1958) 198–225.

[D7] R. Deaux, *Introduction to the Geometry of Complex Numbers*, Ungar, New York, 1956.

[D8] Jesse Douglas, "Edward Kasner," Biographical Memoirs, *National Academy of Sciences*, 31 (1958).

[E1] Howard Eves, *Geometry*, Allyn and Bacon, Boston, 1965.

[G1] P. R. Garabedian, Application of Analytic Continuation to the Solution of Boundary Value Problems, *J. Rational Mechanics and Analysis*, 3 (1954) 383–393.

[G2] ———, *Partial Differential Equations*, Wiley, New York, 1964.

[G3] D. A. Grave, Sur le Problème de Dirichlet, *Assoc. Franc. Bordeaux*, 24 (1895) 111–136.

[G4] Fred Gross, *Factorization of Meromorphic Functions*, Math. Res. Center, Naval Res. Lab., Washington, D. C., 1972 (U. S. Gov. Printing Office).

[G5] H. Grunsky, Flächendifferenzenrechnung in der Funktionentheorie, *Proc. Int. Congr. Math.*, 2 (1954) 114–115.

[G6] ———, Eine funktionentheoretische Integralformel, *Math. Zeitschr.*, 63 (1955) 320–323.

[G7] ———, Ein nichtlineares Randwertproblem im Komplexen, *Math. Nachr.*, 19 (1958) 255–264.

[G8] ———, Über die Umkehrung eines linearen Differentialoperators zweiter Ordnung im Komplexen, *Arkiv der Mathematik*, 14 (1963) 247–251.

[G9] ———, Über die Reduktion eines linearen Differentialoperators zweiter Ordnung im Komplexen, *Math. Ann.*, 163 (1966) 312–320.

[H1] P. Henrici, A Survey of I. N. Vekua's Theory of Elliptic Partial Differential Equations with Analytic Coefficients, *Z. Angew. Math. Phys.*, 8 (1957) 169–203.

[H2] ———, An Algebraic Proof of the Lagrange-Bürmann Formula, *J. Math. Analysis Applications*, 8 (1964) 218–224.

[H3] ———, Fixed Point of Analytic Functions, Report CS 137, Stanford University, July 1969.

[H4] E. Hille, *Analytic Function Theory*, 2 volumes, Ginn, Boston, 1959.

[H5] W. Hurewicz, *Lectures on Ordinary Differential Equations*, Wiley, New York, 1958.

[J1] E. Jabotinsky, Sur les fonctions inverses, *C.R. Acad. Sci. Paris*, 32 (1949) 508–509.

[J2] ———, Representation of Functions by Matrices. Application to Faber Polynomials, *Proc. Amer. Math. Soc.*, 4 (1953) 546–553.

[J3]   E. Jacobsthal, Zur Theorie der linearen Abbildungen, *S.B. Berlin. Math. Ges.*, 33 (1934) 15–34.

[K1]   E. Kasner, Conformal Geometry, *Proc. Fifth. Int. Cong. Math.*, Cambridge, 1912, vol. 2, p. 81.

[K2]   ———, Conformal Classification of Analytic Arcs or Elements, *Trans. Amer. Math. Soc.*, 16 (1915) 333–349.

[K3]   ———, Geometry of Conformal Symmetry (Schwarzian Reflection), *Annals of Math.*, 38 (1937) 873–879.

[K4]   E. Kasner and J. De Cicco, Families of Curves Conformally Equivalent to Circles, *Trans. Amer. Math. Soc.*, 49 (1941) 378–391.

[K5]   Jovan D. Keckic, Analytic and C-Analytic Functions, *Publications de l'Institut Mathématique* (Beograd) N.S., 9 (23) (1969) 189–198.

[K6]   G. Koenigs, Recherches sur les intégrales de certaines équations fonctionnelles, *Annales de l'École Norm. Sup.*, Suppl., 3 (1884) 3–41.

[K7]   M. Kuczma, *Functional Equations in a Single Variable*, Polish Scientific Publishers, Warsaw, 1968.

[L1]   O. Lehto and K. I. Virtanen, *Quasikonforme Abbildungen*, Springer, Berlin, 1965.

[L2]   J. A. Lewis and H. O. Pollak, Photoelastic Calculations by a Complex Variable Method, *ZAMP*, 12 (1961) 30–37.

[L3]   H. Lewy, On the Reflection Laws of Second Order Differential Equations in Two Independent Variables, *Bull. Amer. Math. Soc.*, 65 (1959) 37–58.

[M1]   Z. A. Melzak, Infinite Packings of Disks, *Canad. J. Math.*, 18 (1966) 838–852.

[M2]   Karl Menger, Algebra of Analysis, *Notre Dame Mathematical Lectures*, no. 3, Notre Dame, Ind. (1944).

[M3]   L. M. Milne-Thompson, *Theoretical Hydrodynamics*, 4th ed., Macmillan, New York, 1960.

[M4]   P. Montel, *Leçons sur les récurrences et leurs applications*, Gauthier-Villars, Paris, 1957.

[M5]   Frank Morley and F. V. Morley, *Inversive Geometry*, Ginn, Boston, 1933.

[M6]   J. Moser, On Invariant Curves of Area-Preserving Mappings of an Annulus, *Nachr. Akad. Wiss. Göttingen Math. Phys. Kl. II*, 25 (1961) 1–20.

[N1] Z. Nehari, *Conformal Mapping*, McGraw-Hill, New York, 1952.

[N2] P. F. Neményi and A. W. Sáenz, On the Geometry of Two Dimensional Elastic Stress Fields, *J. Rational Mech. and Analysis*, 1 (1952) 73–86.

[O1] A. Ostrowski, *Solution of Equations and Systems of Equations*, Academic Press, New York, 1960.

[P1] Boyd Patterson, The Differential Invariants of Inversive Geometry, *Amer. J. Math.*, 50 (1928) 553–568.

[P2] D. Pedoe, *Circles*, Pergamon Press, London, 1957.

[P3] G. A. Pfeiffer, On the Conformal Geometry of Analytic Arcs, *Amer. J. Math.*, 37 (1915) 395–430.

[P4] ———, Existence of Divergent Solutions of the Functional Equations, etc., *Bull. Amer. Math. Soc.*, 22 (1916) 163.

[P5] ———, On the Conformal Mapping of Curvilinear Angles, *Trans. Amer. Math. Soc.*, 18 (1917) 185–198.

[P6] E. G. Phillips, *Some Topics in Complex Analysis*, Pergamon Press, London, 1966.

[P7] G. Pólya, On an Integral Function of an Integral Function, *J. London Math. Soc.*, 1 (1926) 12–15.

[R1] L. B. Rall, *Computational Solution of Nonlinear Operator Equations*, Wiley, New York, 1969.

[S1] G. Sansone and J. Gerretsen, *Lectures on the Theory of Functions of a Complex Variable*, Wolters-Noordhoff, Groningen, 1969.

[S2] I. J. Schoenberg, *Approximations: Theory and Practice, Notes*, University of Pennsylvania, 1955.

[S3] I. Schur, On Faber Polynomials, *Amer. J. Math.*, 67 (1945) 33–41.

[S4] H. A. Schwarz, *Mathematische Abhandlungen*, vol. II, Berlin, 1890, p. 151.

[S5] H. Schwerdtfeger, *Geometry of Complex Numbers*, University of Toronto Press, Toronto, 1962.

[S6] C. L. Siegel and J. K. Moser, *Lectures on Celestial Mechanics*, Springer, Berlin, 1971.

[S7]   C. L. Siegel, *Symplectic Geometry*, Academic Press, New York, 1964.

[S8]   J. M. Sloss, Reflection Laws of Second Order Systems of Elliptic Equations in Two Independent Variables with Constant Coefficients, *Pacific J. Math.*, 24 (1968) 541–575.

[S9]   ———, Global Reflection for a Class of Simple Closed Curves, to appear.

[S10]   P. R. Stein and S. M. Ulam, Nonlinear Transformation Studies on Electronic Computers, *Rozprawy Matematyczne*, Warsaw, 39 (1964).

[S11]   E. Study, *Vorlesungen über ausgewählte Gegenstände der Geometrie*, vol. I., Teubner, Leipzig, 1911.

[T1]   J. Traub, *Iterative Methods for the Solution of Equations*, Prentice-Hall, Englewood Cliffs, N.J. 1964.

[V1]   I. N. Vekua, *Generalized Analytic Functions*, Pergamon Press, London, 1962.

[V2]   ———, *New Methods for Solving Elliptic Equations*, North-Holland, Amsterdam—New York, 1967.

[W1]   J. L. Walsh, *Interpolation and Approximation*, American Math. Soc., New York, 1935.

[W2]   O. Wesler, An Infinite Packing Theorem for Spheres, *Proc. Amer. Math. Soc.*, 11 (1960) 324–326.

[W3]   L. C. Woods, Analytic Function Theory, part 5 of *Continuum Physics*, vol. I, A. C. Eringen, ed., Academic Press, New York, 1971.

[Y1]   I. M. Yaglom, *Complex Numbers in Geometry*, Academic Press, New York, 1968.

INDEX